T0260776

An Introduction to Correspondence Analysis

Wiley Series in Probability and Statistics

Established by *Walter A. Shewhart and Samuel S. Wilks*

Editors: *David J. Balding, Noel A. C. Cressie, Garrett M. Fitzmaurice, Geof H. Givens, Harvey Goldstein, Geert Molenberghs, David W. Scott, Adrian F. M. Smith, Ruey S. Tsay*

Editors Emeriti: *J. Stuart Hunter, Iain M. Johnstone, Joseph B. Kadane, Jozef L. Teugels*

The **Wiley Series in Probability and Statistics** is well established and authoritative. It covers many topics of current research interest in both pure and applied statistics and probability theory. Written by leading statisticians and institutions, the titles span both state-of-the-art developments in the field and classical methods.

Reflecting the wide range of current research in statistics, the series encompasses applied, methodological and theoretical statistics, ranging from applications and new techniques made possible by advances in computerized practice to rigorous treatment of theoretical approaches.

This series provides essential and invaluable reading for all statisticians, whether in academia, industry, government, or research.

A complete list of titles in this series can be found at
http://www.wiley.com/go/wsps

An Introduction to Correspondence Analysis

Eric J. Beh
School of Mathematical & Physical Sciences,
University of Newcastle, Australia

Rosaria Lombardo
Department of Economics,
University of Campania "Luigi Vanvitelli," Italy

Registered Offices
John Wiley & Sons, Inc., 111 River Street, Hoboken, NJ 07030, USA
John Wiley & Sons Ltd, The Atrium, Southern Gate, Chichester, West Sussex, PO19 8SQ, UK

Editorial Office
9600 Garsington Road, Oxford, OX4 2DQ, UK

For details of our global editorial offices, customer services, and more information about Wiley products visit us at www.wiley.com.

Wiley also publishes its books in a variety of electronic formats and by print-on-demand. Some content that appears in standard print versions of this book may not be available in other formats.

Library of Congress Cataloging-in-Publication Data

Names: Beh, Eric J., author. | Lombardo, Rosaria, author.
Title: An introduction to correspondence analysis / Eric J. Beh, Rosaria
    Lombardo.
Description: Hoboken, NJ : Wiley, 2021. | Includes bibliographical
    references and index.
Identifiers: LCCN 2020034475 (print) | LCCN 2020034476 (ebook) | ISBN
    9781119041948 (cloth) | ISBN 9781119041962 (adobe pdf) | ISBN
    9781119041979 (epub)
Subjects: LCSH: Correspondence analysis (Statistics)
Classification: LCC QA278.5 .B43 2021 (print) | LCC QA278.5 (ebook) | DDC
    519.5/37–dc23
LC record available at https://lccn.loc.gov/2020034475
LC ebook record available at https://lccn.loc.gov/2020034476

Cover Design: Wiley
Cover Image: © Giovanna Lombardo, p.zza Giovanni XXIII, Castellammare di Stabia (NA) Italy

Set in 9.5/12.5pt STIXTwoText by SPi Global, Chennai, India

Printed and bound by CPI Group (UK) Ltd, Croydon, CR0 4YY

C9781119041948_120321

To Rosey and Alex …
To Donato, Renato and Andrea …

… for your patience, support and always being there

… Eric J. Beh & Rosaria Lombardo

In memory of two pioneers

Jean-Paul Benzecri (1932–2019)

and

John Clifford Gower (1930–2019)

May your legacy live on

Contents

Preface

In the late 2000's we embarked on a rather ambitious project to write a book that covered an extensive array of topics on correspondence analysis. This work resulted in the publication in 2014 of *Correspondence Analysis: Theory, Practice and New Strategies*. The attempt in that book was to provide a comprehensive technical, computational, theoretical and practical description of a variety of correspondence analysis techniques. These focused largely on the analysis of nominal and ordinal categorical variables with a symmetric and asymmetric association structure. We not only described these techniques for two variables but also discussed how they can be used and adapted for analysing multiple categorical variable.

Irrespective of the benefits and faults of that book, we attempted to give an extensive number of different perspectives. While our general flavour may be more in line with the French approach to correspondence analysis we also tried to approach our discussion by incorporating the British/American conventions of categorical data analysis commonly seen throughout the world. A priority we had was to not just provide a synthesis of a broad amount of the correspondence analysis literature from all around the world but to also discuss the role that the origin of categorical data analysis had on the development of correspondence analysis.

From writing the 2014 book we quickly realised that it may contain too much information for someone who was not well versed in some of the more subtle or obscure aspects of correspondence analysis. We also became aware that some didn't feel the need to wade through an extensive literature review and technical discussion, but instead wished to focus on the key features of the analysis. So, after taking some time to take a deep breath and stretch our collective muscles, we dived back into writing again to focus on a book with more of an introductory, or tutorial, flavour than the first book allowed. This book is the result of those deep breaths and muscle stretches.

There are many contributions in the statistics, and allied, literature that provide an introduction to correspondence analysis. However many of these focus primarily on the classical approaches that have been around for decades and deal, for the most part, with the visual depiction of the association between nominal variables. Many of these contributions are also discipline specific so that the terminology used, and the application made, are in terms of a particular data or area of research. Michael Greenacre's book *Correspondence Analysis in Practice*, which is now in its third edition (as of 2017) provides an excellent introductory description of correspondence analysis. Despite the excellent discussion of a wide range of topics, his book focuses on nominal categorical variables and so deals with the more traditional approaches to performing correspondence analysis.

What makes this book distinctive is that we don't just introduce how to perform correspondence analysis for two or more nominal categorical variables using the traditional techniques. This book also provides some introductory remarks on the theory and application of non-symmetrical correspondence analysis; a variant that accommodates for a predictor variable and a response variable. We also provide an introduction to how ordered categorical variables can be incorporated into the analysis. For the analysis of multiple nominal categorical variables we do give an introduction to the classical approaches to multiple correspondence analysis (which involve transforming a multi-way contingency table into a two-way form) but we also provide some introductory remarks and an application of multi-way correspondence analysis; a technique which preserves the *hyper-cube* format of a multi-way contingency table. For the sake of simplicity though, we restrict our attention to the analysis of three variables, but we do examine how to analyse their association when two of them are treated as a predictor variable and the third variable is treated as a response variable.

The one omission from this book is that we do not discuss multi-way correspondence analysis when some of the variables are ordered and some of them are nominal. We could certainly have included some introductory notes on how to perform such an analysis but we felt it was slightly beyond the scope of what we wanted to achieve with this introductory book.

Another point that we would like to make concerns the use of two terms we have used; *association* and *interaction*. We use *association* to refer to the general relationship that exists between two (or more) categorical *variables* while *interaction* describes the relationship between two (or more) specific *categories* from two (or more) variables.

Therefore, this book aims to help researchers improve their familiarity with the concepts, terminology and application of several variants of correspondence analysis. We do describe the theory underlying the statistical and more visual aspects of the analysis. We have also tried to make sure that this book reaches out to students and educators who wish to learn (and teach) the fundamentals of correspondence analysis. In particular, the introductory nature of this book should enable students enrolled in an honours degree (for those in countries such as, but not confined to, Australia, New Zealand, United Kingdom, Canada, Hong Kong and India), a masters program or a PhD in all fields of research to gain appreciation of this form of analysis. The only requirement we ask of the reader is to have some knowledge of introductory statistics. To help the reader, all of the techniques we described can be performed using three R packages that are available on the CRAN; these packages are CAvariants, MCAvariants and CA3variants. However, to avoid an overly long book, we have not provided any guidance on how these packages may be used (reckonizing that they will be constantly updated), although one may refer to their help files for guidance and insight into their use.

So, sit back, relax, and we hope you enjoy your journey into the world of correspondence analysis.

April 2020

Eric J. Beh
Newcastle, Australia
Rosaria Lombardo
Capua, Italy

1

Introduction

1.1 Data Visualisation

Every statistical technique has a long and interesting history. Studying how to numerically and graphically analyse the association between categorical variables is no exception. The contributions of some of the most influential statisticians, including Karl Pearson, R.A. Fisher and G.U. Yule, have left an indelible imprint on how categorical data analysis is performed. Excellent descriptions on the historical development of categorical data analysis, in particular the analysis of contingency tables, can be found by referring to, for example, Goodman and Kruskal (1954) and Agresti (2002, Chapter 16). The influence of the early pioneers has led to almost countless statistical techniques that measure, model, visualise and further scrutinise how categorical variables are related to each other. Much of the key focus has been on the numerical assessment of the strength of the association between the variables – whether the analysis is concerned with two, three or more variables. Yule and Kendall (1950), Bishop et al. (1975) and Liebetrau (1983) also provided excellent discussions of a large number of measures of association for contingency tables. The most influential and widely adopted statistical technique for analysing the association between categorical variables is Pearson's chi-squared statistic (Pearson 1904). The importance and wide applicability of this statistic has been discussed vigourously throughout the literature – see, for example, Lancaster (1969) and Greenwood and Nikulin (1996). The statistic, simply put, is defined as

$$X^2 = \sum \frac{(\text{Observed} - \text{Expected})^2}{\text{Expected}}$$

where "Observed" refers to the observed count made in each cell of a table and "Expected" is its expected value under some model (even if that model reflects independence between the variables). While this statistic can detect if there is a statistically significant association between the variables it does not say anything more about the structure of the association. Various techniques may be considered for examining exactly how the association is structured. These include simple measures such as the product moment correlation (Pearson 1895) which will not only determine the strength of the association but also its direction. Model based approaches such as log-linear models and logistic models are commonly taught as a means of numerically assessing the nature of the association.

An Introduction to Correspondence Analysis, First Edition. Eric J. Beh and Rosaria Lombardo.

Despite the importance of modelling in statistics and her allied fields, there are two issues that need to be considered. Firstly, elementary statistics courses worldwide teach students about the importance of visualising the structure of the data as a means of "seeing" what it looks like before resorting to inferential techniques; this might be through constructing a bar chart, histogram or boxplot of the data. However, in practice many statistical categorical analysis techniques (of course, not all) ignore this visual component altogether and go straight to modelling the structure. Secondly, modelling techniques rely on methodological assumptions of the data, or the perceived behaviour of the data by the analyst. Such thoughts are elegantly, and simply, captured in George Box's (1979) famous quote

> All models are wrong but some are useful

Earlier, Box (1976) had said

> Since all models are wrong, the scientist cannot obtain a "correct" one by excessive elaboration.

Of course, such general phrases have caused a stir amongst the statistical community since a model can never fully capture the "truth" of a phenomenon. We certainly see many advantages in the wide range, and flexibility, of models that are now available but we urge caution when adopting some of them.

An alternative philosophy that can be adopted for assessing the association between the variables of a contingency table is to explore how they are associated to each other by *visualising* the association. There is now a plethora of strategies available for visualising numerical and categorical data. Some of the more popular approaches include the mosaic plot (Friendly 2000, 2002; Theus 2012), the four-fold display (Fienberg 1975) and the cobweb diagram (Upton 2000). The interested reader may also refer to Gabriel (2002) and Wegman and Solka (2002) for the visualisation of multivariate data. The key features of any graphical summary are that what is produced is simple, easy to interpret, and provides a quick and accurate visual representation of data. Cook and Weisberg (1999, p. 29) say of any graphical summary

> In statistical graphics, information is contained in observable shapes and patterns. The task of the creator of a graph is to construct an informative view of the data that is appropriately grounded in a statistical context. The task of the viewer is to find the patterns, and then to interpret their meaning in the same context. Just as an interpretation of a painting or drawing requires understanding of the artist's context, interpreting a graph requires an understanding of the statistical context that surrounds the graph. As in art, conclusions about a graph without understanding the context are likely to be wrong, or off the point at best.

A very good example of the interplay between data visualisation and statistical context can be found by considering Anscombe's quartet (Anscombe 1973). While discussing his point in terms of simple linear regression, Anscombe (1973) provided a compelling argument for the need to visualise data by highlighting four very different scatterplots with equal correlations and equal parameter estimates from a simple linear regression model. His argument shows that the context of the statistical technique needs to be made in terms of the data being analysed, and a visualisation of this context can help the analyst to better understand the statistical and practical contexts of the data being analysed.

1.2 Correspondence Analysis in a "Nutshell"

So where does correspondence analysis fit into this discussion? It is first important to recognise that often the first task in assessing the association structure between categorical variables is to either model, or measure this association, with the structure reflected in the sign and magnitude of a numerical measure. However, as we shall explore in this book, correspondence analysis (in a nutshell) provides a way to visualise the association between two or more categorical variables that form a contingency table. In doing so we gain an understanding of how particular categories from the same variable, or from different variables, "correspond" to each other. From such visual summaries, one can better understand how the variables (and categories) under inspection are associated. In doing so, the analyst can then refine their research question and postulate other structures that may exist in the data. This is all undertaken without the need to make any assumption about the structure of the data, nor does one need to impose untestable, unnecessary, or unnecessarily complicated assumptions on the data (or on the technique). The analyst, whether they are of a technical or practical persuasion, need not rely on a suite of numbers to interpret the association between the variables (unless they want to of course). Therefore, correspondence analysis is a technique that allows the data to inform the analyst of what it is trying to say rather than the model defining how the structure may be defined. The philosophy of letting the "data speak for itself" in correspondence analysis harks back to Jean-Paul Benzécri and his team at the University of Paris, France. Thus, Benzécri is considered to be the father of correspondence analysis although, in truth, many of the technical (and not visual) features stem back to earlier times. Since the early work of Benzécri and his team, the development of correspondence analysis and its many variants has been dominant in many parts of the European statistical, and allied, communities. This is especially so in France, Italy, The Netherlands and Spain. Outside of Europe, it has developed due to the contribution of researchers in Great Britain, Japan and, to a lesser extent, the USA. Unfortunately, in the Australasian region, correspondence analysis has not received the same level of attention as other parts of the world.

Before we continue with our discussion of correspondence analysis, it is worth highlighting that there are many excellent texts on its historical, computational, practical and theoretical development. The first major work that helped to expose correspondence analysis to the English speaking/reading statistical world was that of Hill (1974). Interestingly, he titled his paper "Correspondence analysis: A neglected multivariate method" which was published in the *Journal of the Royal Statistical Society, Series C (Applied Statistics)*. Since then, the growth of correspondence analysis has been quite slow but further insight was made 10 years later with the publication of a book by Michael Greenacre. This book, titled *Theory and Applications of Correspondence Analysis* was published by Academic Press and remains the most cited book of all on the topic and brought correspondence analysis out of the (mainly) French statistical literature and exposed it to the vast English reading/speaking research community; it is thus considered a landmark publication in correspondence analysis. Another excellent book is that of Lebart et al. (1984). Other books that describe the various technical and practical issues of correspondence analysis include, but are definitely not limited to, Greenacre and Blasius (2006), Greenacre (2017), Weller and Romney (1990), Gifi (1990), Benzécri (1992), Clausen (1998), Le Roux and Rouanet (2004), Murtagh (2005), Nishisato (2007) and Kroonenberg (2008). A more recent, technical and historical overview of the variety of correspondence analysis techniques can be found in Beh and Lombardo (2014).

1.3 Data Sets

To describe correspondence analysis, and its key features, we shall be studying its application to a number of contingency tables from a variety of disciplines. In fact, much of the popularity of correspondence analysis rests with its application, not with its technical development. To attest to this, a generic title/abstract/keyword search for "correspondence analysis" on Scopus yields (as of 29 September 2020) 12385 articles concerned with correspondence analysis from the year 2000 onwards. Most recently, 900 publications can be found for the year 2019, in 2018 this number was 832, while 2017 saw 836 publications including this phrase. The most cited article was that of Ter Braak (1986) with 4281 citations followed by Hill and Gauch Jr (1980) with 2805 citations. Both these articles are written with biological/ecological researchers in mind and each propose a variant of the classical approaches to correspondence analysis technique; although we shall not be discussing these variants here.

1.3.1 Traditional European Food Data

Consider the study undertaken by Guerrero et al. (2010) and examined by Beh et al. (2011). The data stems from a study undertaken to see how the word "traditional" (from a food perspective) was perceived across six regions in six European countries; Flanders in Belgium, Burgundy (Dijon) in France, Lazio in Italy, Akershus and Ostfold in Norway, Mazovia (Warsaw) in Poland and Catalonia in Spain. There were two variables of interest in their study. The first was the *Country* where a participant originated from. The second variable, defined here as *Free-word*, consists of a list of 28 words that the recipients were asked to freely associate (that is, no prompting was given) with how they perceived traditional European food; this list was created from more complete list consisting of 1743 valid words. Note that we shall dispense with the quotation marks (" ") around "traditional" for the remainder of this book.

 The data is summarised in Table 1.1 and is based on the six histograms in Figure 1 of Guerrero et al. (2010). They also give an excellent and comprehensive description of the methods used to collect the data and the correspondence analysis that was originally performed using the data from their histograms.

Table 1.1 Cross-classification of *Country* of origin of the participants and the 28 most common *Free-words* that they associate with traditional European food.

Code	Free-word	Belgium	France	Italy	Norway	Poland	Spain	Total
1	Ancient	0	0	2	0	0	15	17
2	Christmas	0	1	0	22	11	9	43
3	Cooking	2	17	0	0	5	9	33
4	Country	0	4	0	17	12	6	39
5	Culture	4	3	0	0	1	7	15
6	Dinner	0	0	0	3	14	0	17
7	Dish	0	10	0	0	11	0	21
8	Family	3	16	5	10	16	13	63
9	Feast	8	9	0	2	0	0	19
10	Good	1	2	6	5	2	12	28
11	Grandmother	10	4	1	5	2	9	31
12	Habit	14	3	0	3	0	25	45
13	Healthy	4	1	3	2	11	6	27
14	Holidays	8	0	0	0	12	1	21
15	Home	1	1	0	0	5	12	19
16	Home-made	0	0	8	2	12	9	31
17	Kitchen	4	0	2	1	4	2	13
18	Meal	1	6	1	0	11	2	21
19	Natural	2	2	5	2	5	5	21
20	Old	2	8	0	6	11	22	49
21	Old-fashioned	10	0	0	5	0	0	15
22	Quality	3	4	0	1	0	4	12
23	Recipe	0	7	0	0	6	0	13
24	Regional	1	7	4	2	1	1	16
25	Restaurant	5	5	1	2	1	1	15
26	Rural	0	12	1	19	5	11	48
27	Simple	2	1	3	1	3	1	11
28	Tasty	8	3	4	0	17	8	40
	Total	93	126	46	110	178	190	743

Source: Based on Guerrero, L., Claret, A., Verbeke, W., Enderli, G., Zakowska-Biemans, S., and Vanhonacker, F. (2010). Perception of traditional food products in six European countries using free word association. Food Quality and Preference, 21:225–233.

Table 1.2 *Temperature* by *Month* in La Guardia airport, 1973.

Temperature (F)	Month					
	May	June	July	August	September	Total
(56, 72]	24	3	0	1	10	38
(72, 79]	5	15	2	9	10	41
(79, 85]	1	7	19	7	5	39
(85, 97]	0	5	10	14	5	34
Total	30	30	31	31	30	152

1.3.2 Temperature Data

Chambers et al. (1983) consider a range of data sets including meteorological data that were collected from the New York State Department of Conservation and the National Wildlife Service. Readings were recorded of ozone, solar radiation, wind and temperature over 152 consecutive days; the collection period was 1 May 1973 to 30 September 1973 (inclusive). Here we shall focus on the association between *Temperature* (measured in Fahrenheit at La Guadia airport, New York) and *Month* (May, June, July, August and September).

The data summarised in Table 1.2 were obtained from the data file `airquality` that is one of the many default data sets in R. The only variation to the data that we have made when cross-classifying the two variables is to replace the numerical labels that were originally given to each month of the study with the name of the month. Refer to Chapter 4 for more details on the analysis of this data.

1.3.3 Shoplifting Data

Consider the shoplifting data summarised in Table 1.3 which contains, in part, the results of a survey undertaken by the Dutch Central Bureau of Statistics (Israëls 1987). They were obtained from a sample of 20819 males who were suspected of shoplifting in Dutch stores between 1977 and 1978. Table 1.3 is a cross-classification of the *Item* stolen (the row variable) and the *Age* groups of the perpetrators (the column variable).

One may treat the association between *Age* and *Item* to be asymmetric such that *Age* is treated as the predictor variable, consisting of 13 categories, and *Item* is the response variable consisting of nine categories. The items categories are *clothing, accessories, tobacco and/or provisions, stationary, books, records, household goods, candy, toys, jewelry, perfume, hobby and/or tools* and *other items*; note that the items listed here have been given labels in Table 1.3 that are in condensed form. The predictor (column) variable consists of the following nine age groups (in years) of the male perpetrators; less than 12 years (< 12), 12 to 14 years (13), 15 to 17 years (16), 18 to 20 years (19), 21 to 29 years (25), 30 to 39 years (35), 40 to 49 years (45), 50 to 64 years (57) and at least 65 years (65+).

We shall be investigating how the *Age* of the perpetrators impacts upon their preference for shoplifting each *Item*; such an association structure is deemed to be *asymmetric*.

Table 1.3 Men's shoplifting data: *Item* stolen versus mid-point of the perpetrator's *Age* interval.

Item Stolen	Age									Total
	< 12	13	16	19	25	35	45	57	65+	
clothing	81	138	304	384	942	359	178	137	45	2568
accessories	66	204	193	149	297	109	53	68	28	1167
tobacco	150	340	229	151	313	136	121	171	145	1756
stationary	667	1409	527	84	92	36	36	37	17	2905
books	67	259	258	146	251	96	48	56	41	1222
records	24	272	368	141	167	67	29	27	7	1102
household	47	117	98	61	193	75	50	55	29	725
candy	430	637	246	40	30	11	5	17	28	1444
toys	743	684	116	13	16	16	6	3	8	1605
jewelry	132	408	298	71	130	31	14	11	10	1105
perfume	32	57	61	52	111	54	41	50	28	486
hobby	197	547	402	138	280	200	152	211	111	2238
other	209	550	454	252	624	195	88	90	34	2496
Total	2845	5622	3554	1682	3446	1385	821	933	531	20819

Section 1.4 provides an interpretation of categorical variables having an *asymmetric association*. We shall also be examining further the nature of this association using a variant of correspondence analysis in Chapter 5 called *singly ordered non-symmetric correspondence analysis*.

1.3.4 Alligator Data

For an exploration of some of the issues concerned with the application of correspondence analysis for multivariate categorical data, we shall confine our attention (for the sake of simplicity) to studying the association among three categorical variables. The data we shall examine is summarised in the $2 \times 5 \times 4$ contingency table of Table 1.4 and comes from Agresti (2002, p. 270). It cross-classifies the *Size* of 219 alligators, their primary *Food* of choice (found in the alligator's stomach) and the Floridian *Lake* in which they reside. The four lakes are Lakes Hancock, Oklawaha, Trafford and George. This data originally came from a study undertaken by the Florida Game and Fresh Water Fish Commission and Agresti (1997, p. 270) notes that this data came from an unpublished manuscript. The interested reader may also refer to Delany and Abercrombie (1986) for earlier data related to a similar study undertaken between 1981 and 1983 from three different lakes that are within 28 km of Gainsville, Florida. The size of the alligators studied in Table 1.4 has been labelled *large* (for those exceeding 2.3 metres in length) and *small* (for those that are no more than 2.3 metres in length). Agresti (1997, p. 268) notes that the five food choices

Table 1.4 Cross-classification of 219 alligators by their *Size*, primary *Food* of choice and *Lake* of residence. Source: Modified from Agresti, A. (2002). Categorical Data Analysis (2nd ed). Wiley, New York.

Size	Fish	Invertebrate	Reptile	Bird	Other
			Primary *Food* of Choice		
			Hancock		
Small	23	4	2	2	8
Large	7	0	1	3	5
			Oklawaha		
Small	5	11	1	0	3
Large	13	8	6	1	0
			Trafford		
Small	5	11	2	1	5
Large	8	7	6	3	5
			George		
Small	16	19	1	2	3
Large	17	1	0	1	3

include the following. For alligators whose primary choice of food was defined as being *invertebrate*, they include those whose stomach contained apple snails, aquatic insects and crayfish. For those whose preferred food choice was *reptile*, their stomachs primarily contained turtles while the stomach of one alligator contained tags of 23 baby alligators that were released the previous year. A category *bird* was defined for those alligators whose stomachs contained the remains of bird wildlife while the category *other* refers to the alligators whose primary choice of food consisted of amphibians, mammals, plant material, stones or other debris or no food at all.

1.4 Symmetrical Versus Asymmetrical Association

Throughout this book we shall be examining two different types of association structure for our visualisation of the association between categorical variables. The distinction we make is on whether the categorical variables are *symmetrically* or *asymmetrically* associated. When using these two terms, we shall treat their association structure to be defined as follows:

- *Symmetrical* association exists when all variables of the contingency table are treated as a predictor variable so that none of them are considered to be a response variable. This definition applies to the analysis of the association for contingency tables formed by cross-classifying two or more categorical variables and is the most commonly adopted type of association structure considered in the analysis of contingency tables.

- *Asymmetrical* association exists when one variable is treated as a response to a second (or multiple) variable(s) that is (are) treated as a predictor variable. While most statistical, and practical, treatments of categorical data treat two such variables as having a symmetrical association there may be practical, or intuitive, reasons for treating the variables as exhibiting an asymmetrical association.

We will adopt these definitions of symmetrical and asymmetrical association by keeping in mind the contingency tables discussed above. For example, a symmetrical association of *Country* and *Free-word* in Table 1.1 considers that neither is treated as a response variable of the other. However if one were to assume that it is known in which *Country* an individual resides, then this information can be used to assess how *Country* impacts upon their perception of traditional European food. Therefore, we may treat the association between the two variables of Table 1.1 as being asymmetrical. In such a case, *Country* is treated as a predictor variable while *Free-word* is treated as the response variable. When quantifying the association between variables that are treated as being symmetrically associated, Pearson's chi-squared statistic is commonly used and is the most appropriate measure to consider. However, when studying the asymmetrical association between two variables alternative measures of association must be used. Obviously, with two different association structures that we are considering in the following chapters, this will impact not only how the association between the variables is to be quantified but it also defines the type of correspondence analysis that will be performed.

For the three-way contingency table of Table 1.4 there are a variety of ways in which an analyst can study the association among the three variables. For example, we may consider the following scenarios:

1. Suppose we know from which *Lake* an alligator comes, and measuring its length determines its *Size*. Given this information we can determine the most likely primary *Food* of choice of the alligator without cutting open its stomach. Of course, this implies not knowing the local wildlife near the lake. It may also be likely that the eco-system at the lake is varied and consists of an array of fish, invertebrates, reptiles, birds and other sources of food.

2. If it is known in which *Lake* an alligator resides and the type of food that dominates this area of Florida, the analyst may then be interested in understanding how these two factors impact on the *Size* of the alligator.

3. Suppose that a rogue alligator is found and it is unknown which lake it comes from but an investigation reveals what type of food was found in its stomach. Then, given the alligators *Size* and primary *Food* of choice, these may be used to determine which *Lake* it is most likely to have come from.

4. Perhaps there is no knowledge of how the three variables are related. So rather than treating the variables as being asymmetrically associated, the analyst may decide the more conservative approach is to analyse the variables symmetrically.

In correspondence analysis, the association structure that is most typically considered is one where the variables are symmetrically associated. The analysis of categorical variables that have an asymmetrical association is only now starting to gain broad international attention. We shall now consider some simple measures of association for two or three categorical variables where the association is treated symmetrically and asymmetrically. We shall first consider such measures for two variables then explore measures for three variables. In the case of multiple categorical variables, the quantities that we describe may be extended easily – although the reader should be aware that the asymmetrical association among more than three categorical variables has not been widely studied in the correspondence analysis literature.

1.5 Notation

To help describe some of the key features of correspondence analysis we now provide an overview of the notation that will be adopted throughout this book. Here, we describe the notation for a two-way contingency table and for a three-way contingency table. Generalisations to multi-way contingency tables are a very straightforward extension to this notation and so, to keep our description as simple as possible, we will not be discussing the notation for these tables any further in this book.

1.5.1 The Two-way Contingency Table

Consider two categorical variables X_1 and X_2 where X_1 consists of I categories and X_2 consists of J categories. We shall denote the I categories of X by $X_{11}, X_{12}, \ldots, X_{1I}$ and the J categories of X_2 by $X_{21}, X_{22}, \ldots, X_{2J}$. If a random sample of n individuals/items is selected then the number of individuals/items in this sample that respond to any category in X_1 and X_2 may be cross-classified. That is, such numbers, or counts, may be summarised in the form of a contingency table which consist of I rows and J columns.

By defining the variables in this way, our attention is focused on analysing the association structure (if it exists) in an I by J, or equivalently $I \times J$, (two-way) contingency table. Such a contingency table will be referred to by **N** where the (i, j)th cell is denoted by n_{ij} for $i = 1, \ldots, I$ and $j = 1, \ldots, J$ and is the joint frequency of individuals/items that are cross-classified into row category X_{1i} and column category X_{2j}. Let the grand total of **N** be n so that the proportion of the sample that is classified into the (i, j)th cell is $p_{ij} = n_{ij}/n$. Therefore

$$\sum_{i=1}^{I} \sum_{j=1}^{J} p_{ij} = 1 .$$

We also define the ith row, and jth column, marginal proportions by

$$p_{i\bullet} = \sum_{j=1}^{J} p_{ij} , \qquad p_{\bullet j} = \sum_{i=1}^{I} p_{ij} ,$$

respectively. Table 1.5 provides a description of the notation of the joint, and marginal, frequencies (or counts) of an $I \times J$ contingency table.

Table 1.5 Notation used for a $I \times J$ contingency table.

X_1/X_2	X_{21}	X_{22}	$\cdots$	X_{2j}	$\cdots$	X_{2J}	Total
X_{11}	n_{11}	n_{12}	$\cdots$	n_{1j}	$\cdots$	n_{1J}	$n_{1\bullet}$
X_{12}	n_{21}	n_{22}	$\cdots$	n_{2j}	$\cdots$	n_{2J}	$n_{2\bullet}$
$\vdots$	$\vdots$	$\vdots$	$\ddots$	$\vdots$	$\ddots$	$\vdots$	$\vdots$
X_{1i}	n_{i1}	n_{i2}	$\cdots$	n_{ij}	$\cdots$	n_{iJ}	$n_{i\bullet}$
$\vdots$	$\vdots$	$\vdots$	$\ddots$	$\vdots$	$\ddots$	$\vdots$	$\vdots$
X_{1I}	n_{I1}	n_{I2}	$\cdots$	n_{Ij}	$\cdots$	n_{IJ}	$n_{I\bullet}$
Total	$n_{\bullet1}$	$n_{\bullet2}$	$\cdots$	$n_{\bullet j}$	$\cdots$	$n_{\bullet J}$	n

1.5.2 The Three-way Contingency Table

Consider now the case where the sample of size n is collected from three categorical variables, X_1, X_2 and X_3. These variables are cross-classified to form a three-way contingency table that consists of I rows, J columns and K categories of the third variable. In the literature, the term given to the third variable of a three-way table is often defined as a *tube*; see, for example Kroonenberg (2008, p. 29) who also notes that if there were a fourth variable a not-so-common term for it is a *pipe* variable. Throughout this book, especially in Chapters 6 and 7, we shall refer to the third variable as consisting of *tube* categories.

We define **N** to be the three-way table of size $I \times J \times K$, that consists of the cell frequencies (or counts). The cell frequency concerned with the ith row, jth column and kth tube – commonly referred to as the (i, j, k)th cell – is denoted by n_{ijk}. Therefore, the proportion of n that are cross-classified into this cell is $p_{ijk} = n_{ijk}/n$ so that

$$\sum_{i=1}^{I}\sum_{j=1}^{J}\sum_{k=1}^{K} p_{ijk} = 1 \ .$$

We also define

$$p_{i\bullet\bullet} = \sum_{j=1}^{J}\sum_{k=1}^{K} p_{ijk} \ , \quad p_{\bullet j\bullet} = \sum_{i=1}^{I}\sum_{k=1}^{K} p_{ijk} \ , \quad p_{\bullet\bullet k} = \sum_{i=1}^{I}\sum_{j=1}^{J} p_{ijk}$$

to be the marginal proportions for the ith row, jth column and kth tube, respectively. Similarly, we define

$$p_{ij\bullet} = \sum_{k=1}^{K} p_{ijk} \ , \quad p_{i\bullet k} = \sum_{j=1}^{J} p_{ijk} \ , \quad p_{\bullet jk} = \sum_{i=1}^{I} p_{ijk}$$

to be the joint marginal proportions by aggregating over the kth tube, jth column and ith row, respectively.

1.6 Formal Test of Symmetrical Association

1.6.1 Test of Independence for Two-way Contingency Tables

Now that we have defined the notation to be used in this book, we shall discuss the various ways in which departures from independence can be described. We will be doing so by confining our attention to the case where two categorical variables are symmetrically associated, and move on to how to amend this approach for asymmetrically associated categorical variables.

Many analysts are very well aware of how to gauge the extent to which the two categorical variables are symmetrically associated, although the term *symmetrically associated* may not always be a phrase that comes to mind. For the (i, j)th cell of a two-way contingency table, symmetrical association is commonly assessed by comparing the known cell frequency n_{ij} with the cell's expected frequency under the hypothesis of independence, $n_{i\bullet}n_{\bullet j}/n$. Therefore, when testing whether there is a statistically significant association (or dependence) between two categorical variables, we typically write the null hypothesis as

$$H_0 : \ n_{ij} = \frac{n_{i\bullet}n_{\bullet j}}{n} \ .$$

This null hypothesis reflects *complete* independence for all $i = 1, \ldots, I$ and $j = 1, \ldots, J$ and can also be stated in terms of the proportions by

$$H_0 : \ p_{ij} = p_{i\bullet}p_{\bullet j} \ .$$

There are other ways in which this hypothesis can be expressed. Three such interrelated ways are described as follows. Firstly, there is

$$H_0 : \ p_{ij} - p_{i\bullet}p_{\bullet j} = 0$$

where the term on the left hand side of the equal sign is the original definition of *contingency* that Pearson described; although note that we describe this contingency in terms of proportions not counts. Secondly, we may define independence between two categorical variables under the null hypothesis of independence by

$$H_0 : \ \frac{p_{ij}}{p_{i\bullet}p_{\bullet j}} = 1 \ .$$

The term on the left hand side of the equal sign is referred to as the *Pearson's ratio* and is just the ratio of the observed cell proportion to its expected value. A third way to express the null hypothesis as

$$H_0 : \ \frac{p_{ij}}{p_{i\bullet}p_{\bullet j}} - 1 = 0 \tag{1.1}$$

where the term on the left hand side of the equal sign is referred to here as *Pearson residual* of the (i, j)th cell of $\mathbf{N}$. All of these hypotheses state the same thing – independence between the two categorical variables. To formally test whether there is a statistically significant association between the variables of a two-way contingency table (at the α level of significance) the most common statistic that is used is Pearson's chi-squared statistic. For categorical variables that have an asymmetrical association structure we shall consider an alternative statistic. We shall discuss this alternative statistic shortly.

1.6.2   The Chi-squared Statistic for a Two-way Table

When testing the null hypothesis that there is complete independence between two categorical variables against its alternative hypothesis that there is dependence (or association), Pearson's chi-squared statistic is used and is defined in terms of the cell and marginal frequencies by

$$X^2 = \sum_{i=1}^{I}\sum_{j=1}^{J} \frac{\left(n_{ij} - \frac{n_{i\bullet}n_{\bullet j}}{n}\right)^2}{\frac{n_{i\bullet}n_{\bullet j}}{n}}. \tag{1.2}$$

For such a test of independence, this statistic is compared with the $1 - \alpha$ percentile of the chi-squared distribution with $(I - 1)(J - 1)$ degrees of freedom.

The chi-squared statistic can also be expressed in terms of proportions. Since the proportion of individuals/units cross-classified into the (i, j)th cell of the contingency table is $p_{ij} = n_{ij}/n$ and the ith row and jth column marginal proportion is $p_{i\bullet} = n_{i\bullet}/n$ and $p_{\bullet j} = n_{\bullet j}/n$ respectively, then Pearson's chi-squared statistic can be expressed as

$$X^2 = n\sum_{i=1}^{I}\sum_{j=1}^{J} \frac{(p_{ij} - p_{i\bullet}p_{\bullet j})^2}{p_{i\bullet}p_{\bullet j}}$$

or, equivalently,

$$X^2 = n\sum_{i=1}^{I}\sum_{j=1}^{J} p_{i\bullet}p_{\bullet j}\left(\frac{p_{ij}}{p_{i\bullet}p_{\bullet j}} - 1\right)^2 \tag{1.3}$$

which is the weighted sum-of-squares of Pearson's residuals. For both these definitions of Pearson's chi-squared statistic, their magnitude is linearly related to the sample size. For example, if the proportions within the contingency table remain unchanged but the sample size is doubled, the chi-squared statistic will also double.

Therefore, it is possible that, even if the marginal proportions remain unchanged, and the sign and magnitude of the "contingencies" do not alter, Pearson's statistic will increase as the sample size increases. The impact of this is that there will always be a sample size for which one can conclude that a statistically significant association exists between the variables, even if the underlying association structure remains unchanged. As we shall see in the coming chapters, correspondence analysis eliminates the impact of the sample size by considering X^2/n as the measure of association between the two categorical variables. This quantity is commonly referred to as the *total inertia* of the contingency table.

1.6.3   Analysis of the Traditional European Food Data

Consider again the traditional European food data summarised in Table 1.1. To gain an understanding of how the different free-words are associated with the six European countries we can perform a chi-squared test of independence using any of the chi-squared statistics defined in the Section 1.6.2. To do so, we first find the expected values of the counts in each cell. Table 1.6 summarises these values to three decimal places. Note that all of the row and column totals of Table 1.6 are equivalent to the marginal totals of Table 1.1.

Table 1.6 Expected cell counts, under independence, for the traditional European food data of Table 1.1.

Code	Free-word	Belgium	France	Italy	Norway	Poland	Spain	Total
					Country			
1	Ancient	2.128	2.883	1.052	2.517	4.073	4.347	17.000
2	Christmas	5.382	7.292	2.662	6.366	10.301	10.994	43.000
3	Cooking	4.131	5.596	2.043	4.886	7.906	8.439	33.000
4	Country	4.882	6.614	2.415	5.774	9.343	9.973	39.000
5	Culture	1.878	2.544	0.929	2.221	3.594	3.836	15.000
6	Dinner	2.128	2.883	1.052	2.517	4.073	4.347	17.000
7	Dish	2.629	3.561	1.300	3.109	5.031	5.370	21.000
8	Family	7.886	10.684	3.900	9.327	15.093	16.110	63.000
9	Feast	2.378	3.222	1.176	2.813	4.552	4.859	19.000
10	Good	3.505	4.748	1.734	4.145	6.708	7.160	28.000
11	Grandmother	3.880	5.257	1.919	4.590	7.427	7.927	31.000
12	Habit	5.633	7.631	2.786	6.662	10.781	11.507	48.000
13	Healthy	3.380	4.579	1.672	3.997	6.468	6.904	28.000
14	Holidays	2.629	3.561	1.300	3.109	5.031	5.370	21.000
15	Home	2.378	3.222	1.176	2.813	4.552	4.859	19.000
16	Home-made	3.880	5.257	1.919	4.590	7.427	7.927	37.000
17	Kitchen	1.627	2.205	0.805	1.925	3.114	3.324	14.000
18	Meal	2.629	3.561	1.300	3.109	5.031	5.370	22.000
19	Natural	2.629	3.561	1.300	3.109	5.031	5.370	21.000
20	Old	6.133	8.310	3.034	7.254	11.739	12.530	49.000
21	Old-fashioned	1.878	2.544	0.929	2.221	3.594	3.836	7.000
22	Quality	1.502	2.035	0.743	1.777	2.875	3.069	12.000
23	Recipe	1.627	2.205	0.805	1.925	3.114	3.324	13.000
24	Regional	2.003	2.713	0.991	2.369	3.833	4.092	16.000
25	Restaurant	1.878	2.544	0.929	2.221	3.594	3.836	15.000
26	Rural	6.008	8.140	2.972	7.106	11.499	12.275	48.000
27	Simple	1.377	1.865	0.681	1.629	2.635	2.813	11.000
28	Tasty	5.007	6.783	2.476	5.922	9.583	10.229	40.000
	Total	93.000	126.000	46.000	110.000	178.000	190.000	743.000

Therefore, using Eq. (1.2), Pearson's chi-squared statistic of Table 1.1 is

$$X^2 = \frac{(0 - 2.128)^2}{2.128} + \frac{(0 - 5.382)^2}{5.382} +, \ldots + \frac{(8 - 10.229)^2}{10.229}$$
$$= 663.76 .$$

Comparing this statistic against the theoretical chi-squared value with $(28 - 1)(6 - 1) = 135$ degrees of freedom gives a p-value that is less than 0.001. Therefore there is ample evidence to conclude that there is a statistically significant association between the 28 different free-words and the European countries listed in Table 1.1. However, the chi-squared test does not determine the nature of the association. That is, the chi-squared test of independence does not identify

1. Whether there is a positive or negative association between the two categorical variables. Although this can be determined by calculating the correlation between *Free-word* and *Country*.
2. How the rows, or columns, are similar or different to each other. Therefore, it is important for us to understand what we mean when we ask "how are the rows similar?" or "how are the columns different?" (say).
3. How particular rows are associated with particular columns, given that a statistically significant association has been found to exist between their variables.

The technique of simple correspondence analysis can provide a visual insight into each of these three issues. The use of the word "simple" does not necessarily imply that correspondence analysis is "simple" or "straightforward". Instead the term "simple" is used to describe the application of correspondence analysis to the most simple of data structures – a two-way contingency table. This technique will be outlined in Chapter 2.

1.6.4 The Chi-squared Statistic for a Three-way Table

Suppose the association between the three categorical variables, X, Y and Z, is symmetrical. The hypothesis of *complete* independence may be specified so that, for the (i, j, k)th cell frequency,

$$H_0 : \quad n_{ijk} = \frac{n_{i\bullet\bullet} n_{\bullet j\bullet} n_{\bullet\bullet k}}{n^2} .$$

We may also examine the *partial* independence or *conditional* independence between the variables of a contingency table. However, for our purposes we shall confine our attention only to the case where we are examining departures from *complete* independence. One may refer to, for example, Agresti (2002) and Loisel and Takane (2016) for more details on *partial* and *conditional* independence.

This null hypothesis may also be expressed in terms of the three-way joint, and marginal, proportions of the contingency table such that

$$H_0 : \quad p_{ijk} = p_{i\bullet\bullet} p_{\bullet j\bullet} p_{\bullet\bullet k}$$

or, equivalently, by the following null hypotheses

$$H_0 : \frac{p_{ijk}}{p_{i\bullet\bullet}p_{\bullet j\bullet}p_{\bullet\bullet k}} = 1 , \tag{1.4}$$

$$H_0 : \frac{p_{ijk}}{p_{i\bullet\bullet}p_{\bullet j\bullet}p_{\bullet\bullet k}} - 1 = 0 . \tag{1.5}$$

Much like the hypotheses described in Section 1.6.1, the left hand side of (1.4) and (1.5) are referred to as the *Pearson ratio* and *Pearson residual* for a three-way contingency table. Formal tests of the statistical significance of the symmetrical association between the variables, under the null hypothesis of *complete* independence, can be undertaken by calculating the three-way chi-squared statistic

$$X^2 = \sum_{i=1}^{I}\sum_{j=1}^{J}\sum_{k=1}^{K} \frac{\left(n_{ijk} - \frac{n_{i\bullet\bullet}n_{\bullet j\bullet}n_{\bullet\bullet k}}{n^2}\right)^2}{\frac{n_{i\bullet\bullet}n_{\bullet j\bullet}n_{\bullet\bullet k}}{n^2}}$$

and testing it against the $1 - \alpha$ percentile of the chi-squared distribution with

$$df = (I - 1)(J - 1) + (I - 1)(K - 1) + (J - 1)(K - 1)$$
$$+ (I - 1)(J - 1)(K - 1)$$

degrees of freedom. Alternatively, but equivalently, the three-way chi-squared statistic can also be expressed in terms of the marginal proportions by

$$X^2 = n\sum_{i=1}^{I}\sum_{j=1}^{J}\sum_{k=1}^{K} \frac{(p_{ijk} - p_{i\bullet\bullet}p_{\bullet j\bullet}p_{\bullet\bullet k})^2}{p_{i\bullet\bullet}p_{\bullet j\bullet}p_{\bullet\bullet k}}$$

and the Pearson residuals by

$$X^2 = n\sum_{i=1}^{I}\sum_{j=1}^{J}\sum_{k=1}^{K} p_{i\bullet\bullet}p_{\bullet j\bullet}p_{\bullet\bullet k}\left(\frac{p_{ijk}}{p_{i\bullet\bullet}p_{\bullet j\bullet}p_{\bullet\bullet k}} - 1\right)^2 . \tag{1.6}$$

1.6.5 Analysis of the Alligator Data

Consider the alligator data of Table 1.4. The chi-squared statistic of this contingency table is 85.492. With

$$df = (2 - 1)(5 - 1) + (2 - 1)(4 - 1)$$
$$+ (5 - 1)(4 - 1) + (2 - 1)(5 - 1)(4 - 1)$$
$$= 4 + 3 + 12 + 12$$
$$= 31$$

degrees of freedom and a p-value that is less than 0.001, there exists a statistically significant association between the *Size* of the alligator, the primary *Food* of choice they consume and the *Lake* in which they reside.

A problem with this statistic is that, on its own, it does not tell us whether there exists an association between just two of the variables, or all pairs of variables or even if there really exists a three-way association. It only tells us that an association exists *somewhere* between the variables. A better picture of the structure of this three-way association can be obtained by partitioning X^2 into three pair-wise chi-squared terms and one that reflects the three-way association between all three variables. We discuss this issue in Chapter 7 before we describe how correspondence analysis can be used to visually assess the nature of the association between three categorical variables.

1.7 Formal Test of Asymmetrical Association

1.7.1 Test of Predictability for Two-way Contingency Tables

The chi-squared statistic is used for testing the statistical significance of the association between categorical variables that are all considered to be predictor variables; that is, when a symmetrical association is of interest. However, in many practical situations, just like with numerical variables, categorical variables may be structured so that some are predictor variables and others are response variables (Takane and Jung 2009a,b). In these cases, Pearson's chi-squared statistic is no longer appropriate to assess the association. When one is studying the association between a single categorical predictor variable and a single categorical response variable, the Goodman–Kruskal tau (Goodman and Kruskal 1954) can be used. Here we describe this statistic which lies at the heart of non-symmetrical correspondence analysis (D'Ambra and Lauro 1989; Lauro and D'Ambra 1984), a technique we outline in Chapter 3.

Suppose we have a column predictor variable and a row response variable. The proportion of individuals/units that are classified into row i given that they belong to column j is given by the ratio $p_{ij}/p_{\bullet j}$. If the row (response) and column (predictor) variables are independent so that $p_{ij} = p_{i\bullet}p_{\bullet j}$, then the unconditional probability of the ith row is $p_{i\bullet}$ and the estimated conditional probability of the ith row (given the column variable) is

$$\frac{p_{ij}}{p_{\bullet j}} = \frac{p_{i\bullet}p_{\bullet j}}{p_{\bullet j}} = p_{i\bullet} \ .$$

Thus, the null hypothesis of zero predictability (knowing something of the columns does not help in our prediction of the rows) is identical to the null hypothesis of complete independence where the hypothesis may be expressed as

$$H_0 : \frac{p_{ij}}{p_{\bullet j}} = p_{i\bullet}$$

or, equivalently,

$$H_0 : \frac{p_{ij}}{p_{\bullet j}} - p_{i\bullet} = 0 \ .$$

We refer to the term on the left hand side of the equal sign as the *Goodman–Kruskal residual* for reasons that will now become apparent.

1.7.2 The Goodman–Kruskal tau Index

Goodman and Kruskal (1954, p. 759) measured the *relative* increase in the predictability of the row variable given the column variable by defining the statistic

$$\tau = \frac{\sum_{i=1}^{I} \sum_{j=1}^{J} p_{\bullet j} \left(\frac{p_{ij}}{p_{\bullet j}} - p_{i\bullet} \right)^2}{1 - \sum_{i=1}^{I} p_{i\bullet}^2} \tag{1.7}$$

where $0 \leq \tau \leq 1$. Eq. (1.7) is referred to as the *Goodman–Kruskal tau index*. An alternative, and equivalent, way to express this index is to let

$$\tau_{\text{num}} = \sum_{i=1}^{I} \sum_{j=1}^{J} p_{\bullet j} \left(\frac{p_{ij}}{p_{\bullet j}} - p_{i\bullet} \right)^2$$

so that

$$\tau = \frac{\tau_{\text{num}}}{1 - \sum_{i=1}^{I} p_{i\bullet}^2} .$$

The numerator of τ, τ_{num}, is the *overall* measure of predictability of the rows given the columns. Note that it is the weighted sum-of-squares of the Goodman–Kruskal residuals where the weights are the marginal proportions of the (given) columns. The denominator measures the overall error in prediction and does not depend on the predictor categories.

When there is (conditional) independence between the rows and columns so that $p_{ij}/p_{\bullet j} - p_{i\bullet} = 0$, there is no relative increase in predictability, and so τ is zero (Agresti 1990). Although, as Agresti (1990, p. 25) describes, low values of τ do not necessarily mean "low" association, since this measure tends to take smaller values as the number of categories increase; see also Kroonenberg and Lombardo (1999). However, Light and Margolin (1971) showed that

$$C = (n - 1)(I - 1)\tau \tag{1.8}$$

is, under the hypothesis of complete independence, asymptotically a chi-squared random variable with $(I - 1)(J - 1)$ degrees of freedom. We shall discuss the Goodman–Kruskal tau index and its role in correspondence analysis some more in Chapter 3.

1.7.3 Analysis of the Traditional European Food Data

Suppose we consider once again the traditional European food data summarised in Table 1.1. Rather than treating the two variables as consisting of predictor categories like we did in Section 1.6.3 we now treat *Country* as being the predictor variable and *Free-word* as being the response variable. That is, we are interested in determining whether the country in which someone resides is a good predictor of how they perceive traditional European food. Doing so, we find that the Goodman–Kruskal tau index is $\tau = 0.035$. While this may appear to be small (since $0 \leq \tau \leq 1$) the C-statistic is

$$C = (747 - 1)(28 - 1) \times 0.035$$

$$= 698.342 .$$

By testing this statistic against $\chi_{0.05}^2$, with $(I - 1)(J - 1) = (28 - 1)(6 - 1) = 135$ degrees of freedom, we find a p-value that is less than 0.001. Therefore, we can conclude that the country in which someone resides is a statistically significant predictor of how they perceive traditional European food. However, this measure does not tell us the countries that are the better predictors and what words are poorly predicted. In Chapter 3 we shall describe how correspondence analysis can be performed to help reveal these asymmetric association structures.

1.7.4 Test of Predictability for Three-way Contingency Tables

Suppose that, for a three-way contingency table, the column and tube variables are treated as predictor variables and the row variable is treated as being the response variable. The proportion of individuals/units that are classified into row i given that they belong to column j and tube k depends on how one assumes the column and tube variables are associated.

Suppose we assume, hypothesise, or it has been determined, that there is complete independence between the two predictor variables. Then we can consider the ratio $p_{ijk}/(p_{\bullet j\bullet}p_{\bullet\bullet k})$. If the three variables are completely independent so that $p_{ijk} = p_{i\bullet\bullet}p_{\bullet j\bullet}p_{\bullet\bullet k}$, then the unconditional probability of the ith row is $p_{i\bullet\bullet}$ and the estimated conditional probability of the ith row (given the jth column and kth tube categories) is

$$\frac{p_{ijk}}{p_{\bullet j\bullet}p_{\bullet\bullet k}} = \frac{p_{i\bullet\bullet}p_{\bullet j\bullet}p_{\bullet\bullet k}}{p_{\bullet j\bullet}p_{\bullet\bullet k}} = p_{i\bullet\bullet} .$$

In this case, the null hypothesis of zero predictability (knowing something of the columns and tubes does not help in our prediction of the rows) is identical to the null hypothesis of complete independence where the hypothesis may be expressed as

$$H_0 : \frac{p_{ijk}}{p_{\bullet j\bullet}p_{\bullet\bullet k}} = p_{i\bullet\bullet}$$

or, equivalently, by

$$H_0 : \frac{p_{ijk}}{p_{\bullet j\bullet}p_{\bullet\bullet k}} - p_{i\bullet\bullet} = 0 .$$

We refer to the term on the left hand side of the equal sign as the *Marcotorchino residual* for reasons that will now become apparent.

1.7.5 Marcotorchino's Index

Marcotorchino's index is a three-way generalisation of the Goodman–Kruskal tau index. However, rather than having a single predictor variable (as the Goodman–Kruskal tau index considers), Marcotorchino's index is a measure of predictability when there exist two predictor variables and a single response variable. For example, for Table 1.4, one may consider that the *Lake* in which an alligator resides and the type of *Food* they eat both impact upon the *Size* of the alligator. Therefore, the aim is to predict the *Size* of alligators, given some knowledge of the type of *Food* they eat and the *Lake* where they can be found.

Marcotorchino's index τ_M, like τ, is a measure of predictability and is defined by

$$\tau_M = \frac{\sum_{i=1}^{I}\sum_{j=1}^{J}\sum_{k=1}^{K}p_{\bullet j\bullet}p_{\bullet\bullet k}\left(\frac{p_{ijk}}{p_{\bullet j\bullet}p_{\bullet\bullet k}} - p_{i\bullet\bullet}\right)^2}{1 - \sum_{i=1}^{I}p_{i\bullet\bullet}^2} . \tag{1.9}$$

Note here that it is hypothesised that there exists complete independence between the two predictor variables, which may or may not be the case. As we shall see in Section 7.5.2, we can assess this independent structure from the partition of τ_M.

The larger the value of this index the higher the power of predictability it has. However, just like the Goodman–Kruskal tau index, a low value of τ_M does not always mean that there is a small quantity of "asymmetric" association between the variables since large numbers of categories will lead to smaller values of the index.

Therefore, to determine the statistical significance of Marcotorchino's index, and following a similar procedure to that outlined by Light and Margolin (1971), the C-statistic

$$C_M = (n-1)(I-1)\tau_M$$

can be calculated for a three-way contingency table; see Beh et al. (2007) for further details. This statistic can then be compared with the $1 - \alpha$ percentile of the chi-squared distribution with

$$\mathrm{df} = (I-1)(J-1) + (I-1)(K-1) + (J-1)(K-1)$$
$$+ (I-1)(J-1)(K-1)$$

degrees of freedom.

1.7.6 Analysis of the Alligator Data

Consider now Agresti's (2002, p. 270) alligator data summarised in Table 1.4. Suppose we treat the row variable as being the response variable and the column and tube variables as being the predictor variables. Therefore, we wish to explore how knowing the *Food* preference of the alligator and the *Lake* in which they are located impact upon their *Size*. We find that the Marcotorchino index for this $2 \times 5 \times 4$ contingency table is $\tau_M = 0.390$. The statistical significance of this index may be assessed by calculating its C-statistic. Doing so, we find that

$$C = (219 - 1)(2 - 1) \times 0.390$$
$$= 85.120 \, .$$

Testing this statistic against the $\chi^2_{0.05}$ value with

$$\mathrm{df} = (2-1)(5-1) + (2-1)(4-1)$$
$$+ (5-1)(4-1) + (2-1)(5-1)(4-1)$$
$$= 4 + 3 + 12 + 12$$
$$= 31$$

degrees of freedom we find that it has a p-value that is less than 0.001. Therefore, the type of *Food* that an alligator eats and the *Lake* in which it may be found are statistically significant predictors of its *Size*. Despite this, the statistic does not reveal exactly how knowing the *Food* preference and *Lake* of the alligator impacts upon its *Size*. The Marcotorchino index, and therefore this C-statistic, imply that *Food* and *Lake* are independent but the statistic does not verify if this is really the case or not. Further information about the asymmetric association between the variables can be obtained if we were to partition τ_M, or even its C-statistic. Doing so provides a numerical summary of how each pair, and all three variables, are associated. More detail about how particular column and tube categories impact upon the outcome of a row category can be obtained by visualising the asymmetric association. These issues will be addressed when we describe multiple correspondence analysis in Chapter 6 and multi-way correspondence analysis in Chapter 7.

1.7.7 The Gray–Williams Index and Delta Index

When studying the asymmetric association between three categorical variables, there are other types of indices that can be considered. Like Marcotorchino's index, these alternative indices can be viewed as three-way analogs of the Goodman–Krusk tau index. Here we will briefly describe two such indices that can be used as the central measure of predictability for the multi-way correspondence analysis of three asymmetrically associated variables. We merely comment on their definition and restrict our description of this variant of correspondence analysis using Marcotorchino's index.

Recall that for Marcotorchino's index, the column and tube predictor variables were assumed to be completely independent. If we relax this assumption then we may consider the ratio $p_{ijk}/p_{\bullet jk}$. In doing so, the null hypothesis of zero predictability is identical to the null hypothesis of complete independence so that

$$H_0 : \quad \frac{p_{ijk}}{p_{\bullet jk}} = p_{i\bullet\bullet}$$

or, equivalently,

$$H_0 : \quad \frac{p_{ijk}}{p_{\bullet jk}} - p_{i\bullet\bullet} = 0 \ .$$

We refer to the term on the left hand side of the equal sign as the *Gray–Williams residual* so that their weighted sum-of-squares is the numerator of the *Gray–Williams index* (Gray and Williams 1981) which is defined by

$$\tau_{GW} = \frac{\sum_{i=1}^{I}\sum_{j=1}^{J}\sum_{k=1}^{K} P_{\bullet jk}\left(\frac{p_{ijk}}{p_{\bullet jk}} - p_{i\bullet\bullet}\right)^2}{1 - \sum_{i=1}^{I}p_{i\bullet\bullet}^2} \ .$$

Yet another possibility that may arise for three asymmetrically associated categorical variables is the case where there is one predictor (tube) variable and two response (row and column) variables (Lombardo, 2011). In this case, and when assuming that the association between the two response variables is assumed to be independent, the null hypothesis of zero predictability is

$$H_0 : \quad \frac{p_{ijk}}{p_{\bullet\bullet k}} = p_{i\bullet\bullet}p_{\bullet j\bullet}$$

or, equivalently,

$$H_0 : \quad \frac{p_{ijk}}{p_{\bullet\bullet k}} - p_{i\bullet\bullet}p_{\bullet j\bullet} = 0 \ .$$

Here, the term on the left hand side of the equal sign is referred to as the *Delta residual* and the weighted sum-of-squares of these residuals gives the numerator of the *Delta index* (Lombardo 2011) which is defined by

$$\tau_D = \frac{\sum_{i=1}^{I}\sum_{j=1}^{J}\sum_{k=1}^{K}P_{\bullet\bullet k}\left(\frac{p_{ijk}}{p_{\bullet\bullet k}} - p_{i\bullet\bullet}p_{\bullet j\bullet}\right)^2}{1 - \sum_{i=1}^{I}p_{i\bullet\bullet}^2} \ . \tag{1.10}$$

Pearson's chi-squared statistic, Marcotorchino's index, the Gray–Williams index and the Delta index can all be used as the central measure of association (or predictability) for the multi-way correspondence analysis of a three-way contingency table. However, for the sake of brevity, in Chapter 7 we shall confine our attention to demonstrating the role of Pearson's and Marcotorchino's measures for the analysis of the association between the three variables of Table 1.4.

1.8   Correspondence Analysis and R

Since the early 2000's there has been a steady increase in the number of packages available on the CRAN for performing correspondence analysis. One of the earliest, and simplest packages is the MASS package. Venables and Ripley (2002), Beh and Lombardo (2014) and Ripley (2016) give an overview of the features and application of this package for performing correspondence analysis. The MASS package provides the analyst with the flexibility to perform either a simple or a multiple correspondence analysis with the option of including supplementary points onto a display; we shall not consider supplementary points in this book but the interested reader can refer to Greenacre (1984), for example, for more information on this issue. The ca package of Nenadić and Greenacre (2007) includes functions for performing simple, multiple and joint correspondence analysis using two and three dimensions for the graphical displays; joint correspondence analysis is a variant of multiple correspondence analysis that we will not discuss in this book. The ca.r function of Murtagh (2005, pp 18–20) can also be considered. Applications based on R packages such as ade4 (Chessel et al. 2004; Dray and Dufour 2007; Thioulouse et al. 1997) and vegan (Oksanen et al. 2016) are also available. The ade4 package is very comprehensive in the variety of correspondence analysis techniques that can be considered and is aimed primarily at researchers in the ecological and environmental disciplines; it can perform such techniques as canonical correspondence analysis, discriminant correspondence analysis and decentred correspondence analysis. Other packages that can be used to perform correspondence analysis include those of anacor (De Leeuw and Mair 2009b), cabootcrs (Ringrose 2012), CAinterprTools (Alberti 2015), cncaGUI (Librero et al. 2020), dualScale (Clavel et al. 2014), ExPosition (Beaton et al. 2014), FactoMineR (Lê et al. 2008), homals (De Leeuw and Mair 2009a) and Ptak (Leibovici 2010, 2015). These packages cover wide a range of varying theoretical and practical issues. For example, the vegan package can be used for constrained correspondence analysis, canonical correspondence analysis and detrended correspondence analysis while the cncaGUI package can be used to perform a canonical correspondence analysis or a canonical non-symmetrical correspondence analysis. Table 1.7 provides an overview of the type of correspondence analysis techniques these R packages perform for two variables, with a focus on those techniques described in this book. Table 1.8 summarises these packages for the types of techniques they describe for multiple variables. The labels attached to each of the techniques are described as follows and the "x" that are in bold in the tables are variants that shall be described in the coming chapters:

- CA – the classical approach to the correspondence analysis of two nominal categorical variables. Chapter 2 will provide an overview of this approach.
- NSCA – the classical approach to the non-symmetrical correspondence analysis of two nominal categorical variables. Chapter 3 will provide an overview of this approach.

Table 1.7 R packages and some variants of correspondence analysis for two variables.

	Variants of Simple Correspondence Analysis					
Package	CA	NSCA	DOCA	SOCA	DONSCA	SONSCA
ade4	x	x				
anacor	x					
ca	x					
cabootcrs	x					
CAinterprTools	x					
CAvariants	x	x	x	x	x	x
dualScale	x					
ExPosition	x					
FactoMineR	x					
homals	x					
MASS	x					
PTAk	x					
vegan	x					

Table 1.8 R packages and some variants of correspondence analysis for multiple variables.

	Variants of Multiple Correspondence Analysis					
Package	MCA	CA3	JCA	NSCA3	OCA3	ONSCA3
ade4	x					
anacor	x					
ca	x		x			
CA3variants		x		x	x	x
cabootcrs	x					
CAinterprTools	x					
CAvariants						
dualScale	x					
ExPosition	x					
FactoMineR	x					
homals	x					
MASS	x					
MCAvariants	x					
PTAk		x				
vegan	x					

- SOCA – singly ordered correspondence analysis examines the association between a nominal categorical variable and an ordinal categorical variable and is described by Beh (2001a) and Lombardo and Beh (2016).
- DOCA – doubly ordered correspondence analysis is a variation of the correspondence analysis of a two-way contingency table where both variables consist of ordinal categories. This approach will be described in Chapter 4 and was originally proposed by Beh (1997).
- DONSCA – a non-symmetrical correspondence analysis technique for a doubly ordered two-way contingency table that was described by Lombardo, Beh and D'Ambra (2007).
- SONSCA – a non-symmetrical correspondence analysis technique for a singly ordered two-way contingency table that was described by Lombardo, Beh and D'ambra (2011). Chapter 5 will provide an overview of some of the key features of this variant.
- MCA – the multiple correspondence analysis of a contingency table consisting of three or more nominal variables and will be described in Chapter 6. It involves transforming a multi-way contingency table into a table consisting of just rows and columns.
- CA3 – a special case of multi-way correspondence analysis that is an alternative to multiple correspondence analysis. In its most general form, this variant involves analysing the data in its "cube" (for three variables – leading to CA3) or "hyper-cube" form (for more than three variables). This type of analysis is confined to the study of the association between nominal variables but may be symmetrically or asymmetrically associated. We briefly describe this variant in Chapter 7.
- OCA3 – a variant of CA3 but where at least one of the variables consists of ordered categories. This variant was described by Beh and Lombardo (2014, Section 10.8) and Lombardo, Beh and Kroonenberg (2020), and deals with the variables being symmetrically associated.
- NSCA3 – a variant of CA3, and a three-way extension of the non-symmetrical correspondence analysis for three nominal and asymmetrically associated categorical variables and was originally described by Lombardo, Carlier and D'ambra (1996). We provide a description of this variant in Chapter 7 by analysing the alligator data summarised in Table 1.4.
- ONSCA3 – a variant of NSCA3 but where at least one of the variables consists of ordered categories; see Lombardo, Beh and Kroonenberg (2020).
- JCA – *joint correspondence analysis* is a variation of MCA described in Greenacre (1988, 1990) and Greenacre and Blasius (2006) and helps to greatly improve the quality of the visual display obtained from performing a MCA.

A more comphrensive description of these packages can be found in Beh and Lombardo (2014, 2019a) and Lombardo and Beh (2016).

The focus of this introductory book is to describe the key features of correspondence analysis for nominal and ordinal categorical variables that are symmetrically and asymmetrically associated. Generally, the packages listed above describe a wide range of correspondence analysis issues for nominal and symmetrically associated variables. However, as Table 1.7 and Table 1.8 suggest, only a few of them address the issue of ordered variables and/or asymmetric association. Despite this, our book performs all the necessary calculations in R using the following three packages

- CAvariants (Version 5.5)
- MCAvariants (Version 2.5),
- CA3variants (Version 2.5).

An overview of the CAvariants package is made by Lombardo and Beh (2016) and is freely available on the Comprehensive R Archive Network (CRAN) from the URL

http://cran.r-project.org/web/packages/CAvariants/index.html

while the MCAvariants and CA3variants packages can be freely downloaded from

http://cran.r-project.org/web/packages/MCAvariants/index.html

and

http://cran.r-project.org/web/packages/CA3variants/index.html

respectively. All three packages provide a range of functionalities for performing simple correspondence analysis, multiple correspondence analysis and three-way correspondence analysis for variables consisting of nominal and/or ordinal categories that are symmetrically or asymmetrically associated.

There are a variety of other computing facilities that can perform correspondence analysis, including the commercial packages JMP, SPSS, XLStat, Minitab and SAS, with varying degrees of flexibility. Non-commercial packages are also available. For example, PAST (PAleonotological STatistics), described by Hammer et al. (2001), can be freely downloaded from //folk.uio.no/ohammer/past/, as can Ludovic Lebart's DtmVic5.6 package from www.dtmvic.com. Pieter Kroonenberg developed the package 3WayPack for performing multi-way correspondence analysis (and related techniques); this may be freely downloaded from http://three-mode.leidenuniv.nl/. Refer to Beh and Lombardo (2014, Chapter 12) for more of a discussion and history (including references therein) on a variety of computational issues concerned with correspondence analysis.

1.9 Overview of the Book

Now that we have briefly described what correspondence analysis is and defined some of the notation that we shall be adopting, we outline the specific correspondence analysis techniques that will be described in this book. There are a huge variety of correspondence analysis techniques that one can perform. For example, Beh and Lombardo (2014) cite 35 variations of correspondence analysis that existed at that time (no doubt there are now more), while Beh and Lombardo (2012, 2019a) provides an extensive genealogy and discussion of many of these variants (we freely admit that there are unintentional exclusions in these discussions). Some of them deal with subtle, and not-so-subtle, variations in the way the data are treated. Some delve into issues that the classical techniques do not address. Many of the techniques are largely confined in specific areas of research and so are tailored to specific data structures. Although, just like every area of research, the growth in the development and application of correspondence analysis is ever increasing (although its growth may not be at the same pace as other statistical techniques). We shall not be considering all of these variations (in fact, in our 2014 book we only provided an overview of some of them). Instead we shall be focusing our attention in this book on the following six techniques:

- simple correspondence analysis
- non-symmetrical correspondence analysis

- ordered simple correspondence analysis
- ordered non-symmetrical correspondence analysis
- multiple correspondence analysis
- multi-way correspondence analysis.

Therefore, this book will describe the key features and application of each of them in the following six chapters that are divided into three parts:

- In Part 1, we present a description of two variants of correspondence analysis that can be applied to a two-way contingency table where the two variables consist of *nominal* categories.

  Chapter 2 explores some of the issues concerned with *simple correspondence analysis*. In this chapter, we introduce how a visual summary of the association between two nominal variables can be obtained, and how to interpret such a visual display. Such displays can be obtained using *singular value decomposition* (SVD) and using its features we will introduce the traditional *correspondence plot*. We then move on to discuss the construction and interpretation of the *biplot*. A demonstration of these issues will be made by analysing the traditional European food data of Table 1.1 including a description of what the two-dimensional displays reveal about the association between the variables *Free-word* and *Country*. We also discuss what these displays do not describe about this association.

  For the classical approach to correspondence analysis described in Chapter 2 it is implied that the association structure between the two variables of the contingency table are *symmetrically* associated. That is, as we described in Section 1.4, both variables are treated as predictor variables. However, there are situations where an asymmetric association may exist, or is assumed to exist. In this case, we can analyse the association using *non-symmetrical correspondence analysis* and we shall focus on this variant in Chapter 3. In doing so, we examine the features of this approach and apply it to Table 1.1 where the *Country* from which a respondent comes is a predictor for how they perceive traditional European food.

- In Part 2 we continue our discussion of simple correspondence analysis and non-symmetrical correspondence analysis but shift our focus from *nominal* categorical variables to the analysis of the association between *ordinal* categorical variables. Interestingly, the vast majority of (but not all) descriptions and applications of correspondence analysis treat, or assume, that ordered variables consist of nominal categories. Chapter 4 demonstrates how the ordered structure of such variables can be incorporated into the analysis. Rather than using SVD, as the classical approach does, when a two-way contingency table consists of two ordered categorical variables we can instead use *bivariate moment decomposition* (BMD). Therefore, Chapter 4 will examine the key features and application of this variant of correspondence analysis by analysing the association between the two variables, *Temperature* and *Month*, of Table 1.2. We then move on to a discussion of the non-symmetrical correspondence analysis of ordered categorical variables in Chapter 5. In particular, this chapter will explore the case where one variable is ordered and the second variable is nominal. For the analysis of this asymmetric association structure, we can use a *hybrid decomposition* (HD) which is a combination of the features from SVD (for the nominal variable) and BMD (for the ordered variable). We shall be performing this variant of correspondence analysis using HD on the shoplifting data of Table 1.3. For this analysis the ordered column categories (*Age* of the perpetrators) will be treated as the predictor variable and the nominal row categories (*Item* stolen) as the response variable.

- Part 3 will extend our discussions in the first two chapters and describe the correspondence analysis of multiple nominal categorical variables. It consists of Chapter 6 which describes *multiple correspondence analysis* while Chapter 7 describes *multi-way correspondence analysis*. Both variants analyse multiple categorical variables, but the way in which they do it is very different. Broadly speaking, multiple correspondence analysis involves transforming a multi-way contingency table into a two-way table and then simple correspondence analysis is applied. The types of transformation include converting the multi-way contingency table into its indicator or Burt matrix forms; these will be discussed in more detail in Chapter 6. One can also stack categories from the same variable and obtain a visual display of the association. We shall be examining the application of these three approaches. A second way that correspondence analysis can be applied to a multi-way contingency is to use multi-way correspondence analysis. Unlike multiple correspondence analysis, multi-way correspondence analysis treats the variables in its "natural" state. That is, it treats the data as a *data-cube* for three variables or as a *hyper-cube* when more than three cross-classified categorical variables are analysed. In doing so, multi-way extensions of SVD can be used and in Chapter 7 we focus our discussion on the role of the *Tucker3 decomposition* (Tucker 1966) for performing multi-way correspondence analysis. In this chapter we provide an outline of how this variant of correspondence analysis can be performed for three symmetrically and asymmetrically associated variables.

The multiple correspondence analysis and multi-way correspondence analysis of ordinal categorical variables is an ongoing area of research and one that is beyond the scope of this introductory book. However, Beh and Lombardo (2014) describe some of the theory and application of the work that is being done in this area. We therefore invite the interested reader to investigate these variants of correspondence analysis at their leisure.

With this sequence of chapters in mind, our discussion of correspondence analysis is aimed at the casual and the more serious user.

Part I

Classical Analysis of Two Categorical Variables

2

Simple Correspondence Analysis

2.1 Introduction

Recently, there has been an explosion in the statistical and allied literature on descriptions of correspondence analysis. Some of these provide an excellent and all-encompassing overview of the technique while others provide a discussion of the most basic features with an applied focus. In this chapter we shall focus our attention on describing the key mathematical and practical features of correspondence analysis for a two-way contingency table. In doing so, the technique we discuss here is referred to as *simple correspondence analysis*. It is hoped that such a discussion is of equal relevance to those interested in exploring some of the theory that lies behind the technique and those who wish to gain more insight into its application. Our discussion will focus on how one may obtain a visual summary of the symmetric association between two nominal categorical variables. As we shall discuss, the way in which the visualisation is achieved is by comparing the distribution of the relative cell frequencies for each row and each column of the table. There are many ways in which this may be performed. Most approaches focus on obtaining row scores and column scores that best describe the variation in the categories while maximising the association between them. Some of these ignore the visual component altogether while others determine such scores in a variety of different ways and highlight their similarities and differences visually. Therefore, there are many ways in which correspondence analysis can be viewed and so many techniques are synonymous with it. These include, but are not limited to, *reciprocal averaging* (Hill 1974), *homogeneity analysis* (Gifi 1990), *principal component analysis of categorical data* (Torgerson 1958), *dual scaling* (Nishisato 1994), *optimal scaling* (Bock 1956), *canonical correlation analysis* (Hirschfeld 1935), *Guttman scaling* (Guttman 1941) and *correspondence factor analysis* (Teil 1975). One may refer to Tenenhaus and Young (1985) for an excellent discussion of the various links that some of these approaches share. Further historical insight into the development of correspondence analysis can be found in Benzécri (1977), van Meter et al. (1994) and Holmes (2008) who provide a very good account of this history from a French perspective. Further historical descriptions may be found in Aramatte (2008), de Falguerolles (2008) and Lebart (1988) for a historical overview of the computational, theoretical and practical development of correspondence analysis. For discussions of these developments for categorical data analysis in general the interested reader may refer to Agresti (2002, pg. 619–631), Bishop et al. (1975), Imrey et al. (1981), Stigler (2002) and Fienberg and Rinaldo (2007). Since correspondence analysis deals extensively with

An Introduction to Correspondence Analysis, First Edition. Eric J. Beh and Rosaria Lombardo.
© 2021 John Wiley & Sons Ltd. Published 2021 by John Wiley & Sons Ltd.

contingency tables, the interested reader is also invited to read Killion and Zahn (1976) who provide an impressive bibliography on the history and development of contingency tables up to 1974.

2.2 Reducing Multi-dimensional Space

2.2.1 Profiles Cloud of Points

Suppose we consider the *i*th row of Table 1.5 which consists of the elements

$$(n_{i1}, n_{i2}, \cdots, n_{ij}, \cdots, n_{iJ}).$$

This sequence of cell frequencies can be thought of as representing a point for the *i*th row in a *J*-dimensional space. Similarly, the *j*th column of Table 1.5

$$(n_{1j}, n_{2j}, \cdots, n_{ij}, \cdots, n_{Ij})$$

can be thought of as a point in an *I*-dimensional space.

For *I* and *J* greater than 3, the graphical representation of these points is not possible, but exists in a multi-dimensional space referred to as a *cloud-of-points*; the row *cloud-of-points* consists of *J* dimensions and the column *cloud-of-points* consists of *I* dimensions. Therefore, the goal is to derive a low-dimensional space consisting of no more than $\min(I, J)$ dimensions that jointly represents the position of, and variation between, the categories of the contingency table. Figure 2.1 provides a visual display of this concept and we shall discuss further how this dimension reduction can be achieved in the following sections. Although, before we discuss this issue we first introduce the idea of a *profile* for each of the rows and columns of a two-way contingency table.

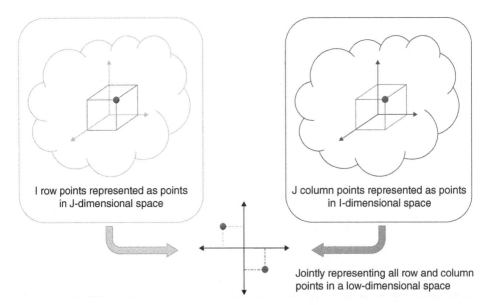

I row points represented as points in J-dimensional space

J column points represented as points in I-dimensional space

Jointly representing all row and column points in a low-dimensional space

Figure 2.1 Reduction of the row and column clouds-of-points to a low-dimensional subspace.

2.2.2 Profiles for the Traditional European Food Data

Suppose we wish to graphically summarise the association between the two variables of the traditional European food data summarised in Table 1.1. Since the contingency table is of size 28 × 6, each row (or category of the variable *Free-word*) can be represented as a point in a cloud-of-points consisting of six dimensions. Similarly, a visual comparison of the six European countries can be undertaken such that each column of Table 1.1 can be viewed in a cloud-of-points consisting of 28 dimensions. Of course, it is impractical to view clouds-of-points that consist of more than three dimensions and so we need to find a way of reducing their dimensionality but still preserve as much as possible the association that exists between the two variables.

To obtain such a visual summary of the association we need to take into account that not all free-words have the same number of classifications; this is apparent because the cell frequencies and the row marginal frequencies of each row of Table 1.1 are not the same. Nor does each country consist of the same number of people sampled. Therefore, we need to find a way to reflect that the row (and column) marginal totals of Table 1.1 are different. This can be done by comparing the relative distribution of the rows of Table 1.1. Instead of comparing the distribution of cell counts for each row we can instead compare their *profiles*. For example, the free-word *Culture* (Code 5) consists of 15 classifications so its profile is defined by the relative distribution of its frequencies so that

$$\left(\frac{4}{15}, \frac{3}{15}, \frac{0}{15}, \frac{0}{15}, \frac{1}{15}, \frac{7}{15}\right) = (0.267, \ 0.200, \ 0.000, \ 0.000, \ 0.067, \ 0.467)$$

and shows that a nearly 50% of responses for this word were from *Spain*. On the other hand, no-one from *Italy* or *Norway* associated *Culture* (Code 5) with traditional European food. This profile is somewhat similar to the profile of the free-word *Habit* (Code 12):

$$\left(\frac{14}{45}, \frac{3}{45}, \frac{0}{45}, \frac{3}{45}, \frac{0}{45}, \frac{25}{45}\right) = (0.311, \ 0.067, \ 0.000, \ 0.067, \ 0.000, \ 0.556) \ .$$

Suppose we now consider respondents who associated traditional European food with *Tasty* (Code 28). Its profile is

$$\left(\frac{8}{40}, \frac{3}{40}, \frac{4}{40}, \frac{0}{40}, \frac{17}{40}, \frac{8}{40}\right) = (0.200, \ 0.075, \ 0.100, \ 0.000, \ 0.425, \ 0.200)$$

and is dominated by those surveyed in *Poland*. We can see that this profile is very different from those we obtained for the free-words *Culture* (Code 5) and *Habit* (Code 12).

A more detailed comparison of the row (*Free-word*) profiles can be made by graphically depicting the row profiles and the column profiles. Figure 2.2 gives a visual summary of the first 10 row profiles. Unfortunately, if we were to plot the profiles of all 28 free-words, the result would be a mess and would not provide an informative, or easy to interpret, comparison of the row profiles. Similarly, we can compare the six column (*Country*) profiles across the 28 free-words. Figure 2.3 provides a comparison of the profiles for all six European countries studied.

2.2.3 Weighted Centred Profiles

On a more general note, the profile of the *i*th row category can be expressed in terms of the proportions of a contingency table (rather than its counts) by

$$\left(\frac{n_{i1}}{n_{i\bullet}}, \frac{n_{i2}}{n_{i\bullet}}, \ \dots \ , \frac{n_{ij}}{n_{i\bullet}}, \ \dots \ , \frac{n_{iJ}}{n_{i\bullet}}\right) = \left(\frac{p_{i1}}{p_{i\bullet}}, \frac{p_{i2}}{p_{i\bullet}}, \ \dots \ , \frac{p_{ij}}{p_{i\bullet}}, \ \dots \ , \frac{p_{iJ}}{p_{i\bullet}}\right)$$

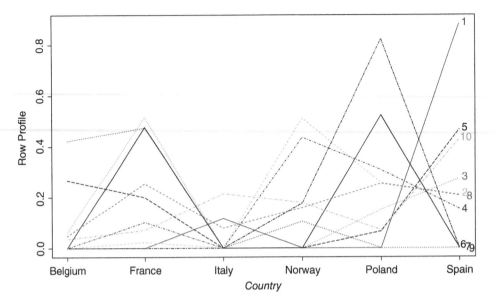

Figure 2.2   The profiles of the first 10 rows of Table 1.1.

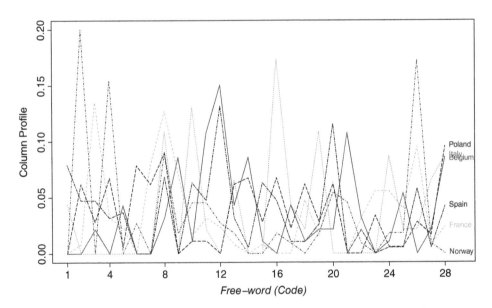

Figure 2.3   Profiles of the six European countries of Table 1.1.

and is termed the *row profile* of the ith row category. Similarly, the *column profile* of the jth column category is

$$\left(\frac{n_{1j}}{n_{\bullet j}}, \frac{n_{2j}}{n_{\bullet j}}, \; \cdots \;, \frac{n_{ij}}{n_{\bullet j}}, \; \cdots \;, \frac{n_{Ij}}{n_{\bullet j}} \right) = \left(\frac{p_{1j}}{p_{\bullet j}}, \frac{p_{2j}}{p_{\bullet j}}, \; \cdots \;, \frac{p_{ij}}{p_{\bullet j}}, \; \cdots \;, \frac{p_{Ij}}{p_{\bullet j}} \right).$$

If there is no association between the row and column variables, such that $p_{ij} = p_{i\bullet}p_{\bullet j}$, for all $i = 1, 2, \ldots, I$ and $j = 1, 2, \ldots, J$, then the ith row and jth column profiles simplify to

$$\left(\frac{p_{i\bullet}p_{\bullet 1}}{p_{i\bullet}}, \frac{p_{i\bullet}p_{\bullet 2}}{p_{i\bullet}}, \; \cdots \;, \frac{p_{i\bullet}p_{\bullet j}}{p_{i\bullet}}, \; \cdots \;, \frac{p_{i\bullet}p_{\bullet J}}{p_{i\bullet}} \right) = (p_{\bullet 1}, p_{\bullet 2}, \; \cdots \;, p_{\bullet j}, \; \cdots \;, p_{\bullet J})$$

and

$$\left(\frac{p_{1\bullet}p_{\bullet j}}{p_{\bullet j}}, \frac{p_{2\bullet}p_{\bullet j}}{p_{\bullet j}}, \; \cdots \;, \frac{p_{i\bullet}p_{\bullet j}}{p_{\bullet j}}, \; \cdots \;, \frac{p_{I\bullet}p_{\bullet j}}{p_{\bullet j}} \right) = (p_{1\bullet}, p_{2\bullet}, \; \cdots \;, p_{i\bullet}, \; \cdots \;, p_{I\bullet}),$$

respectively. So rather than considering the profile of each category, we can instead compare the centred row, and centred column, profiles to detect any departures from independence. In doing so the ith *centred row profile* is

$$\left(\frac{p_{i1}}{p_{i\bullet}} - p_{\bullet 1}, \; \frac{p_{i2}}{p_{i\bullet}} - p_{\bullet 2}, \; \cdots \;, \; \frac{p_{ij}}{p_{i\bullet}} - p_{\bullet j}, \; \cdots \;, \; \frac{p_{iJ}}{p_{i\bullet}} - p_{\bullet J} \right) \tag{2.1}$$

while the jth *centred column profile* is

$$\left(\frac{p_{1j}}{p_{\bullet j}} - p_{1\bullet}, \; \frac{p_{2j}}{p_{\bullet j}} - p_{2\bullet}, \; \cdots \;, \; \frac{p_{ij}}{p_{\bullet j}} - p_{i\bullet}, \; \cdots \;, \; \frac{p_{Ij}}{p_{\bullet j}} - p_{I\bullet} \right).$$

In both cases, if there is complete independence between the row and column categories, both sets of centred profiles will consist of zeros.

These centred profiles can be expressed in terms of Pearson's chi-squared statistic. To show this, consider again the chi-squared statistic defined by Eq. (1.2). It can be expressed as the weighted sum-of-squares of the centred row profiles such that

$$X^2 = n \sum_{i=1}^{I} \sum_{j=1}^{J} \left[\sqrt{\frac{p_{i\bullet}}{p_{\bullet j}}} \left(\frac{p_{ij}}{p_{i\bullet}} - p_{\bullet j} \right) \right]^2.$$

Therefore, by defining

$$r_{j|i} = \sqrt{\frac{p_{i\bullet}}{p_{\bullet j}}} \left(\frac{p_{ij}}{p_{i\bullet}} - p_{\bullet j} \right), \tag{2.2}$$

the chi-squared statistic can be expressed as the sum-of-squares of these values so that

$$X^2 = n \sum_{i=1}^{I} \sum_{j=1}^{J} r_{j|i}^2.$$

In this case

$$\mathbf{r}_i = (r_{1|i}, \; \cdots \;, r_{J|i})$$

$$= \left(\sqrt{\frac{p_{i\bullet}}{p_{\bullet 1}}} \left(\frac{p_{i1}}{p_{i\bullet}} - p_{\bullet 1} \right), \; \cdots \;, \sqrt{\frac{p_{i\bullet}}{p_{\bullet J}}} \left(\frac{p_{iJ}}{p_{i\bullet}} - p_{\bullet J} \right) \right)$$

is the *weighted centred profile* for the ith row category. Similarly, defining

$$c_{i|j} = \sqrt{\frac{P_{\bullet j}}{P_{i\bullet}}} \left(\frac{P_{ij}}{P_{\bullet j}} - P_{i\bullet} \right)$$

means that their sum-of-squares gives the chi-squared statistic so that

$$X^2 = n \sum_{i=1}^{I} \sum_{j=1}^{J} c_{i|j}^2 .$$

In this case

$$\mathbf{c}_j = (c_{1|j}, \ldots, c_{I|j})$$

$$= \left(\sqrt{\frac{P_{\bullet j}}{P_{1\bullet}}} \left(\frac{P_{1j}}{P_{\bullet j}} - P_{1\bullet} \right), \ldots, \sqrt{\frac{P_{\bullet j}}{P_{I\bullet}}} \left(\frac{P_{Ij}}{P_{\bullet j}} - P_{I\bullet} \right) \right)$$

is defined as the *weighted centred profile* for the jth column category. Note that by simply rearranging the $r_{j|i}$ and $c_{i|j}$ gives equivalent results and are just the *Pearson standardised residual*, z_{ij} of the (i, j)th cell, defined as

$$z_{ij} = \frac{P_{ij} - P_{i\bullet}P_{\bullet j}}{\sqrt{P_{i\bullet}P_{\bullet j}}} = r_{j|i} = c_{i|j} . \tag{2.3}$$

The rows of Table 2.1 are the weighted centred row profiles of the contingency table given by Table 1.1. Similarly, the columns of Table 2.1 are the weighted centred column profiles of the contingency table.

Figure 2.4 visually summarises the first ten weighted centred row profiles of Table 1.1; like for those we discussed for Figure 2.2 depicting all 28 row profiles will make the display very cluttered. Similarly, Figure 2.5 visualises the weighted centred column profiles. Even with a relatively small number (six) of column profiles they do not provide a straightforward comparison of the categories within the *Country* variable. What makes viewing Figure 2.4 and Figure 2.5 (including Figure 2.2 and Figure 2.3) even more difficult is that they all provide virtually no helpful insight into the association between each of the 28 words and the six countries studied; recall that this is one of the key goals that we are trying to achieve. However, we can gain some understanding about how the (weighted centred) row profiles differ, or not. Similarly, we can also gain some insight into how the (weighted centred) column profiles categories compare.

A detailed look of the complete set of weighted centred row and column profiles reveals that there are profiles that are similar and there are also those that are different. For example, Figure 2.4 shows that

- *Norway* associates more strongly than any other country the words *Christmas* (Code 2) and *Rural* (Code 26) with traditional European food.
- *France* associates more strongly than any other country the word *Cooking* (Code 3) with traditional European food.
- *Poland* associates more strongly than any other country the word *Dinner* (Code 6) with traditional European food.
- *France* and *Poland* are equally likely, and more so, than any other country to associate *Dish* (Code 7) with traditional European food.

Table 2.1 The Pearson standardised residuals, z_{ij}, for Table 1.1.

		Country					
Code	Free-word	Belgium	France	Italy	Norway	Poland	Spain
1	Ancient	−0.051	−0.062	0.031	−0.060	−0.074	0.188
2	Christmas	−0.081	−0.085	−0.061	0.212	0.009	−0.021
3	Cooking	−0.033	0.177	−0.054	−0.084	−0.037	0.008
4	Country	−0.077	−0.037	−0.059	0.158	0.033	−0.046
5	Culture	0.064	0.011	−0.036	−0.057	−0.050	0.060
6	Dinner	−0.051	−0.062	−0.039	0.007	0.181	−0.076
7	Dish	−0.057	0.126	−0.043	−0.067	0.098	−0.085
8	Family	−0.057	0.060	0.016	<0.001	0.009	−0.028
9	Feast	0.145	0.118	−0.041	−0.022	−0.078	−0.080
10	Good	−0.045	−0.046	0.112	0.009	−0.066	0.067
11	Grandmother	0.126	−0.020	−0.027	0.001	−0.073	0.015
12	Habit	0.134	−0.066	−0.003	−0.062	−0.124	0.134
13	Healthy	0.017	−0.063	0.031	−0.025	0.061	−0.015
14	Holidays	0.133	−0.069	−0.043	−0.067	0.114	−0.069
15	Home	−0.029	−0.045	−0.041	−0.064	0.008	0.119
16	Home-made	−0.075	−0.091	0.131	0.032	0.039	−0.005
17	Kitchen	0.070	−0.056	0.041	−0.006	0.013	−0.030
18	Meal	−0.035	0.044	−0.014	−0.049	0.092	−0.056
19	Natural	−0.009	−0.030	0.113	−0.027	−0.000	−0.005
20	Old	−0.055	−0.003	−0.066	−0.024	−0.007	0.099
21	Old-fashioned	0.049	−0.040	−0.025	0.135	−0.047	−0.049
22	Quality	0.051	0.051	−0.033	−0.024	−0.062	0.020
23	Recipe	−0.045	0.119	−0.034	−0.053	0.060	−0.067
24	Regional	−0.022	0.096	0.105	−0.013	−0.053	−0.056
25	Restaurant	0.092	0.057	0.001	−0.009	−0.050	−0.053
26	Rural	−0.086	0.050	−0.044	0.150	−0.070	−0.013
27	Simple	0.025	−0.023	0.098	−0.021	0.009	−0.039
28	Tasty	0.059	−0.053	0.031	−0.092	0.089	−0.025

These observations are also apparent by "eye-balling" the counts in Table 1.1. By comparing the weighted centred column profiles depicted in Figure 2.5, we can see that the profile of each of the six countries appears quite different. Figure 2.5 also shows that:

- The free-words *Christmas* (Code 2) and *Old-fashioned* (Code 21) have profiles that are quite similar (except for in *Belgium* and *Spain*).
- The centred (and uncentred) profile of the free-words *Dish* (Code 7) and *Recipe* (Code 23) are virtually identical. This suggests that those in the study saw very little difference in the

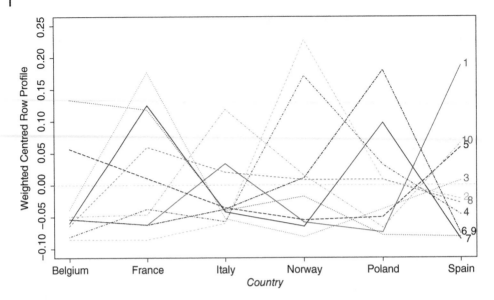

Figure 2.4 Weighted centred row profiles, r_j, of the first 10 rows of Table 1.1.

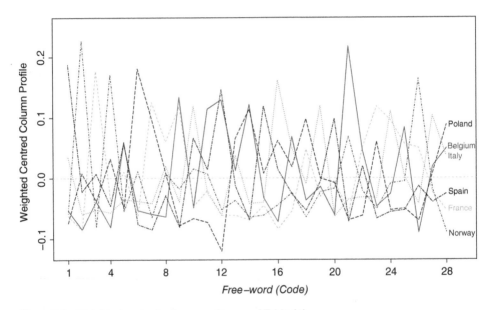

Figure 2.5 Weighted centred column profiles, c_j, of Table 1.1.

interpretation of these two free-words when associating them with the word "traditional", from an European food perspective.

- The profile of the free-words *Culture* (Code 5), *Grandmother* (Code 11) and *Quality* (Code 22) are all very similar.
- The profile of the free-words *Good* (Code 10) and *Home* (Code 15) are very similar.
- The profiles of the free-words *Cooking* (Code 3) and *Restaurant* (Code 25) are very different from one another.

- The profiles of the free-words *Cooking* (Code 3) and *Holidays* (Code 14) are very different from one another.
- The centred profile for the free-words *Natural* (Code 19) and *Old* (Code 20) lie very close to zero. This suggests that all of the countries felt that neither free-word strongly exemplified "traditional" European food. Also, the near-zero values for these two profiles suggest that their expected cell frequencies are consistent with what is expected if the variables *Free-word* and *Country* were independent.

We do wish to emphasise that such a list is not intended to provide an exhaustive comparison of similar and different profiles. Rather, it is intended to highlight a few such cases where it is appropriate for our discussions below.

2.3 Measuring Symmetric Association

2.3.1 The Pearson Ratio

The comparison of the row (and column) profiles forms the foundations of performing simple correspondence analysis. Now that we have defined and discussed these profiles, and their weighted and centred versions, for Table 1.1 we turn our attention to better understanding the nature of the association between these rows and columns. In short, the aim of correspondence analysis is to reflect the nature of the association between categorical variables by determining scores for each category that describe how similar, or different, these profiles are. From these scores we obtain the visual summary of the association that is the key feature of correspondence analysis. So, for a two-way contingency table, the strength of the association between the row scores and column scores should also be measured. This is traditionally done by considering those cell frequencies that deviate from what is expected under the hypothesis of independence. For the (i, j) cell, *complete* independence arises when

$$p_{ij} = p_{i\bullet}p_{\bullet j}$$

for all $i = 1, 2, \ldots, I$ and $j = 1, 2, \ldots, J$. Note that we discussed this point in Section 1.6.1. However, it will rarely happen that complete independence between the row variable and column variable will ever be satisfied. So a multiplicative measure of the departure from complete independence can be considered such that

$$p_{ij} = \gamma_{ij}p_{i\bullet}p_{\bullet j} .$$

In this case, complete independence between the row and column variables will arise when $\gamma_{ij} = 1$ for all $i = 1, 2, \ldots, I$ and $j = 1, 2, \ldots, J$. The γ_{ij} term is referred to as the *Pearson ratio* of the (i, j)th cell; see also Goodman (1996) and Beh (2004b) for such a measure. They define this quantity as the ratio of the observed cell proportion to what is expected under complete independence so that

$$\gamma_{ij} = \frac{p_{ij}}{p_{i\bullet}p_{\bullet j}} . \tag{2.4}$$

A related measure is the *Pearson residual* for the (i, j)th cell of the contingency table and is defined by

$$\gamma_{ij} - 1 = \frac{p_{ij}}{p_{i\bullet}p_{\bullet j}} - 1 \tag{2.5}$$

so that complete independence will be observed when $\gamma_{ij} - 1 = 0$ or, equivalently $\gamma_{ij} = 1$, for all $i = 1, 2, \ldots, I$ and $j = 1, 2, \ldots, J$.

2.3.2 Analysis of the Traditional European Food Data

Consider again the traditional European food data found in Table 1.1. The Pearson residuals, $\gamma_{ij} - 1$, for this table are summarised in Table 2.2.

Table 2.2 The Pearson residuals, $\gamma_{ij} - 1$, for the traditional European food data of Table 1.1.

		Country					
Code	Free-word	Belgium	France	Italy	Norway	Poland	Spain
1	Ancient	−1.000	−1.000	0.900	−1.000	−1.000	2.450
2	Christmas	−1.000	−0.863	−1.000	2.456	0.068	−0.182
3	Cooking	−0.516	2.038	−1.000	−1.000	−0.368	0.067
4	Country	−1.000	−0.395	−1.000	1.944	0.284	−0.398
5	Culture	1.130	0.179	−1.000	−1.000	−0.722	0.825
6	Dinner	−1.000	−1.000	−1.000	0.192	2.438	−1.000
7	Dish	−1.000	1.808	−1.000	−1.000	1.186	−1.000
8	Family	−0.620	0.498	0.282	0.072	0.060	−0.193
9	Feast	2.364	1.793	−1.000	−0.289	−1.000	−1.000
10	Good	−0.715	−0.579	2.461	0.206	−0.702	0.676
11	Grandmother	1.577	−0.239	−0.479	0.089	−0.731	0.135
12	Habit	1.486	−0.607	−1.000	−0.550	−1.000	1.173
13	Healthy	0.184	−0.782	0.795	−0.500	0.701	−0.131
14	Holidays	2.044	−1.000	−1.000	−1.000	1.385	−0.814
15	Home	−0.580	−0.690	−1.000	−1.000	0.098	1.470
16	Home-made	−1.000	−1.000	3.168	−0.564	0.616	0.135
17	Kitchen	1.458	−1.000	1.485	−0.480	0.284	−0.398
18	Meal	−0.620	0.685	−0.231	−1.000	1.186	−0.628
19	Natural	−0.239	−0.438	2.846	−0.357	−0.006	−0.069
20	Old	−0.674	−0.037	−1.000	−0.173	−0.063	0.756
21	Old-fashioned	4.326	−1.000	−1.000	1.252	−1.000	−1.000
22	Quality	0.997	0.966	−1.000	−0.437	−1.000	0.304
23	Recipe	−1.000	2.175	−1.000	−1.000	0.927	−1.000
24	Regional	−0.501	1.580	3.038	−0.156	−0.739	−0.756
25	Restaurant	1.663	0.966	0.077	−0.099	−0.720	−0.739
26	Rural	−1.000	0.474	−0.663	1.674	−0.565	−0.104
27	Simple	0.453	−0.464	3.405	−0.386	0.138	−0.644
28	Tasty	0.598	−0.558	0.615	−1.000	0.774	−0.218

Recall that Pearson's chi-squared statistic for Table 1.1 is 663.76. With a p-value that is less than 0.001, this statistic shows that there is a statistically significant association between the rows and column variables of the contingency table. The Pearson residuals summarised in Table 2.2 show that there are values of $\gamma_{ij} - 1$ that are close to zero; see, for example the free-word *Natural* (Code 19) in *Poland* which has a residual of −0.006. See also the Pearson residual of 0.060 for the free-word *Family* (Code 8) in *Poland*. Such values show there are some observed cell frequencies that are consistent with what is expected if the rows and columns of Table 1.1 were independent. There are also Pearson residuals for the traditional European food data that are a long way from zero; this is due in part to the presence of zero's in Table 1.1 (such cells have a Pearson residual of −1.000). See also the free-word *Old-fashioned* (Code 21) in *Belgium* (with a residual of 4.326) and the free-word *Simple* (Code 27) in *Italy* (with a Pearson residual of 3.405). These values highlight those cell frequencies where the observed number of classifications is greater than what is expected if *Free-word* and *Country* were independent. Hence, these cells suggest possible sources of association between the two variables. However, the Pearson residuals do not formally confirm that such sources are the reason for the statistically significant association between *Free-word* and *Country*, although their weighted sum of squares is related to the chi-squared statistic by

$$X^2 = n \sum_{i=1}^{I} \sum_{j=1}^{J} p_{i\bullet} p_{\bullet j} (\gamma_{ij} - 1)^2 \tag{2.6}$$

which is just Eq. (1.3).

Since we know that there exists a statistically significant association between the two variables of Table 1.1, we can gain some insight into potential sources of association by examining further the magnitude of the Pearson residuals. To do so, we now consider strategies that provide a visual summary of how strongly each row, and column, category contributes to the association structure. We shall show how this can be achieved through decomposing the Pearson residual into components that are used to graphically depict this association.

2.4 Decomposing the Pearson Residual for Nominal Variables

2.4.1 The Generalised SVD of $\gamma_{ij} - 1$

The association that exists between the row and column variables of a two-way contingency table is preserved within each of the Pearson residuals, $\gamma_{ij} - 1$, defined by Eq. (2.5). To maximise the association between the categorical variables using as few dimensions as possible we decompose these residuals. The method of decomposition used depends on the nature of the variables that are being examined. In cases where both the row and column variables consist of nominal categories we shall consider a generalised singular value decomposition (GSVD) of the Pearson residuals. In practice, when an analyst is performing a correspondence analysis on their data, traditionally the categories are treated as being nominal and the conventional approach to correspondence analysis (which we shall outline in the following sections) is performed. When the row and column variables each consist of ordered categories an alternative method of decomposition needs to be considered to reflect this ordinal structure. In this case, Chapter 4 will describe the bivariate moment decomposition (BMD) and its implementation for decomposing the Pearson residuals.

For now we confine our attention to the case where the row and column variables of a two-way contingency table are nominal. In this case the Pearson residuals defined by Eq. (2.5) are decomposed using GSVD such that

$$\gamma_{ij} - 1 = \sum_{s=1}^{S} a_{is} \lambda_s b_{js} . \tag{2.7}$$

The right hand side of Eq. (2.7) is the GSVD of the (i, j)th Pearson residual. The decomposition is designed to obtain row and column scores that are maximally correlated, and in doing so gives a visual summary of the association that uses as few dimensions as possible. For the analysis of two nominal categorical variables, the maximum number of dimensions needed to visually summarise the association is $S = \min(I, J) - 1$. Typically, the plot commonly used in correspondence analysis to visually summarise the association between the variables that consists of exactly S dimensions is referred to as the *optimal correspondence plot*; a low-dimensional space, or sub-space, of this optimal space is simply referred to as a *correspondence plot*. For each dimension of the correspondence plot, a *score* is determined for each row and column category. From Eq. (2.7), the score for the ith row on the sth dimension (for $i = 1, 2, \dots, I$ and $s = 1, 2, \dots, S$) is denoted by a_{is}. These row scores are better known in the correspondence analysis literature as the elements of the *left generalised vector* and are constrained so that

$$\sum_{i=1}^{I} p_{i\bullet} a_{is} = 0, \qquad \sum_{i=1}^{I} p_{i\bullet} a_{is}^2 = 1 \tag{2.8}$$

for each of the S dimensions. Similarly, the score of the jth column on the sth dimension of the S-dimensional visual display is denoted by b_{js} (for $j = 1, 2, \dots, J$ and $s = 1, 2, \dots, S$). These column scores are better known in the correspondence analysis literature as the elements of the *right generalised vector* and are constrained so that

$$\sum_{j=1}^{J} p_{\bullet j} b_{js} = 0, \qquad \sum_{j=1}^{J} p_{\bullet j} b_{js}^2 = 1 \tag{2.9}$$

for each of the S dimensions. From a statistical perspective, these constraints imply that the expectation and variance of a_{is}, say, along the sth dimension is

$$E(a_{is}) = \sum_{i=1}^{I} p_{i\bullet} a_{is}$$
$$= 0$$

and

$$Var(a_{is}) = E(a_{is}^2) - [E(a_{is})]^2$$
$$= \sum_{i=1}^{I} p_{i\bullet} a_{is}^2 - 0$$
$$= 1 .$$

Similarly, the expectation and variance of b_{js} along the sth dimension is zero and one, respectively. For more information on this issue, Beh and Simonetti (2011) provide an interpretation of higher order moments of related quantities from a correspondence analysis perspective. Their focus was on non-symmetrical correspondence analysis, a variant of the correspondence analysis that we shall describe in Chapter 3.

The correlation of the row scores and the column scores along the sth dimension of the optimal (S-dimensional) sub-space is

$$
\begin{aligned}
\lambda_s &= \mathrm{Corr}(a_{is},\ b_{js}) \\
&= \sum_{i=1}^{I}\sum_{j=1}^{J} p_{ij} \frac{(a_{is} - \mathrm{E}(a_{is}))}{\sqrt{\mathrm{Var}(a_{is})}} \frac{(b_{js} - \mathrm{E}(b_{js}))}{\sqrt{\mathrm{Var}(b_{js})}} \\
&= \sum_{i=1}^{I}\sum_{j=1}^{J} p_{ij} a_{is} b_{js}
\end{aligned}
\tag{2.10}
$$

by virtue of the above constraints imposed on a_{is} and b_{js}. Here λ_s is termed the sth *singular value* of the set of Pearson's residuals. The choice of which dimension is helpful for visualising the association is made through the property that the singular values are arranged in descending order so that the first singular value, λ_1 provides the "highest" correlation between the row scores and the column scores. As such, the ith row score and the jth column score is denoted by a_{i1} and b_{j1}, respectively. Similarly, the next "highest" correlation is $\lambda_2 < \lambda_1$ with scores for the ith row and jth column of a_{i2} and b_{j2}, respectively. Generally, the singular values are arranged so that

$$
1 = \lambda_0 > \lambda_1 > \lambda_2 > \dots > \lambda_S > 0 .
$$

One may also note that the sth singular value of the matrix of Pearson's residuals can be expressed in terms of Pearson ratio's such that

$$
\lambda_s = \sum_{i=1}^{I}\sum_{j=1}^{J} \gamma_{ij} p_{i\bullet} p_{\bullet j} a_{is} b_{js}
$$

so that where there is complete independence between the row and column variables such that $\gamma_{ij} = 1$ for all $i = 1, \dots, I$ and $j = 1, \dots, J$ then

$$
\begin{aligned}
\lambda_s &= \sum_{i=1}^{I}\sum_{j=1}^{J} \gamma_{ij} p_{i\bullet} p_{\bullet j} a_{is} b_{js} \\
&= \left(\sum_{i=1}^{I} p_{i\bullet} a_{is} \right) \left(\sum_{j=1}^{J} p_{\bullet j} b_{js} \right) \\
&= 0
\end{aligned}
$$

for all $s = 1, \dots, S$.

An important feature of the singular values is that Pearson's chi-squared statistic can be partitioned such that

$$
X^2 = n \sum_{s=1}^{S} \lambda_s^2 .
\tag{2.11}
$$

Therefore, if all of the singular values are zero then the association between the rows and columns of the contingency table is consistent with what is expected under the hypothesis of complete independence between them. Also, large values of λ_s reflect a large value of the Pearson chi-squared statistic, although a large sample size n can also lead to a large statistic (even if all λ_s are considered small). Therefore, to remove the impact that the sample size has on the magnitude of the Pearson chi-squared statistic, X^2/n is the preferred measure of association in correspondence analysis. This quantity is referred to as the *total inertia* of the contingency table.

It is important to note that the GSVD of the Pearson residuals is one way to perform a correspondence analysis on a contingency table. There are alternative measures that can be decomposed with different weights attached to the a_{is} and b_{js} values. The approach we have taken here is to treat the foundation of correspondence analysis from the perspective of considering $\gamma_{ij} - 1$ as the central measure of association for the (i, j)th cell of a two-way contingency table. One may instead consider z_{ij} as this measure; see Eq. (2.3). Alternative, and yet equivalent, solutions involve a more analytic, or even geometric, approach to find the same set of solutions to λ_s, a_{is} and b_{js}. We shall leave it for the interested reader to consider these alternative strategies at their leisure.

2.4.2 SVD of the Pearson Ratio's

While Section 2.4.1 described correspondence analysis by applying a GSVD to the Pearson residuals, $\gamma_{ij} - 1$, an equivalent strategy is to apply a decomposition to the weighted centred row and column profiles. While Eq. (2.3) showed that the elements of these profiles is equivalent to Pearson's standardised residual, z_{ij}, the row and column scores (a_{is} and b_{js}, respectively) along the sth dimension of a visual display can be obtained through the SVD of these residuals

$$z_{ij} = \frac{p_{ij} - p_{i\bullet}p_{\bullet j}}{\sqrt{p_{i\bullet}p_{\bullet j}}} = \sum_{s=1}^{S} \tilde{a}_{is} \lambda_s \tilde{b}_{js} \; .$$

Here $\tilde{a}_{is}$ and $\tilde{b}_{js}$ are constrained by

$$\sum_{i=1}^{I} \tilde{a}_{is} = 0, \qquad \sum_{i=1}^{I} \tilde{a}_{is}^2 = 1$$

and

$$\sum_{j=1}^{I} \tilde{b}_{js} = 0, \qquad \sum_{j=1}^{J} \tilde{b}_{js}^2 = 1 \; ,$$

respectively, and λ_s is defined by the correlation between $\tilde{a}_{is}$ and $\tilde{b}_{js}$.

2.4.3 GSVD and the Traditional European Food Data

To demonstrate the features that come from the GSVD of the Pearson residuals of the traditional European food data in Table 1.1, consider its Pearson residuals summarised in Table 2.2. The numerical features of the GSVD of Table 2.2 are given as follows. Table 2.3 gives the a_{is} values from this decomposition and it can be easily verified in R that they satisfy

Table 2.3 The a_{is} values from the GSVD of $\gamma_{ij} - 1$ for Table 1.1.

Code	Free-word	Dim 1	Dim 2	Dim 3	Dim 4	Dim 5
1	Ancient	1.349	-2.104	1.675	-1.495	-0.644
2	Christmas	−0.977	−1.542	−1.468	0.768	−0.221
3	Cooking	0.013	1.162	−0.271	−2.029	−0.375
4	Country	−1.132	−0.930	−1.267	0.535	−0.261
5	Culture	1.501	0.251	0.146	−0.564	−0.835
6	Dinner	−2.197	0.367	0.609	1.779	−1.663
7	Dish	−1.518	1.899	0.100	−0.921	−0.737
8	Family	−0.483	0.128	−0.066	−0.466	0.328
9	Feast	1.223	2.110	−1.584	−0.265	0.828
10	Good	0.280	−1.299	0.749	−0.424	1.535
11	Grandmother	1.255	0.109	−0.551	0.467	0.085
12	Habit	1.960	−0.471	0.045	−0.062	−0.884
13	Healthy	−0.208	0.082	0.945	0.884	−0.054
14	Holidays	0.194	1.429	0.630	2.288	−1.295
15	Home	0.548	−0.870	1.116	−0.646	−1.888
16	Home-made	−0.653	−0.684	1.899	0.410	1.277
17	Kitchen	0.540	0.374	0.781	1.541	0.804
18	Meal	−1.093	1.213	0.693	−0.148	−0.620
19	Natural	−0.099	−0.277	1.131	0.216	1.715
20	Old	0.077	−0.548	0.145	−0.697	−1.105
21	Old-fashioned	2.141	0.744	−2.014	2.798	0.789
22	Quality	1.196	0.593	−0.676	−0.916	−0.131
23	Recipe	−1.405	2.013	−0.100	−1.282	−0.525
24	Regional	−0.326	0.740	0.037	−1.221	3.018
25	Restaurant	0.845	1.268	−0.901	0.064	1.090
26	Rural	−0.567	−0.800	−1.474	−0.606	0.368
27	Simple	−0.084	0.271	1.094	0.862	2.333
28	Tasty	−0.002	0.601	1.092	0.871	−0.220

the constraints given by Eq. (2.8). Note there are $I = 28$ rows (for each free-word) and $S = 5$ columns (for each dimension of the optimal correspondence plot). Similarly Table 2.4 summarises the b_{js} values from the GSVD and satisfies Eq. (2.9). There are $J = 6$ rows in Table 2.4 (one for each of the six countries of Table 1.1) and $S = 5$ columns (again, for each dimension of the optimal correspondence plot). The $S = 5$ squared singular values of the Pearson residuals of Table 1.1 are

$$\lambda_1^2 = 0.237, \quad \lambda_2^2 = 0.205, \quad \lambda_3^2 = 0.185, \quad \lambda_4^2 = 0.164, \quad \lambda_5^2 = 0.101$$

Table 2.4 The b_{js} values from the GSVD of $\gamma_{ij} - 1$ for Table 1.1.

Academic Level	Dim 1	Dim 2	Dim 3	Dim 4	Dim 5
Belgium	1.891	1.126	−0.409	1.402	0.127
France	−0.277	1.292	−0.680	−1.605	0.333
Italy	−0.198	−0.444	1.934	0.171	3.338
Norway	−0.654	−1.240	−1.784	0.598	0.500
Poland	−1.159	0.467	0.701	0.747	−0.750
Spain	0.771	−1.020	0.560	−0.709	−0.678

and are arranged in descending order. Multiplying the sum-of-squares of these singular values by the sample size, n, gives the chi-squared statistic of Table 1.1 so that

$$X^2 = 743 \times (0.237 + 0.205 + 0.185 + 0.164 + 0.101)$$
$$= 663.759$$

thereby verifying the link between the chi-squared statistic of the contingency table and their singular values; see Eq. (2.11).

From a correspondence analysis perspective, the importance of each of the terms from the GSVD of the Pearson residuals, $\gamma_{ij} - 1$, that is summarised in Table 2.2 is that they can be used to visually, and numerically, describe the association between the variables *Country* and *Free-word*. Therefore, we shall now turn our attention to describing how the values of a_{is}, b_{js} and λ_s can be used to graphically detect similarities, and differences, between the row profiles and the column profiles.

2.5 Constructing a Low-Dimensional Display

2.5.1 Standard Coordinates

Recall that the purpose of applying the GSVD to the set of Pearson residuals is to reduce the dimensionality of the two clouds-of-points to a single low-dimensional sub-space (such as the optimal correspondence plot) while also maximising the amount of association that exists between the variables; see Figure 2.1. For the traditional European food data of Table 1.1, one approach to constructing a low-dimensional plot, and one that may be enticing to the analyst due to its simplicity, is to use the I rows of Table 2.3 as the coordinates for the points in a low-dimensional plot consisting of no more than $S = 5$ dimensions; recall that Table 2.3 summarises the a_{is} values. Similarly, we could also use each row of Table 2.4 to define the coordinates of the points of each column in this plot; that is, use the b_{js} values as coordinates for the jth column category on the sth dimension. When such points are used to visualise the association between the row and column categories, they are referred to as *standard coordinates*.

However, there are two problems with using standard coordinates to visually summarise the association between the row and column categories of Table 1.1. Firstly, they are defined in such a way that the association structure between the variables has not yet been captured. This is because the row scores, a_{is}, and the column scores, b_{js}, do not reflect the association between the variables; the correlation between them, λ_s, ensures that it is maximised although it is not incorporated into the calculation of the standard coordinates. Instead these scores can be used to identify free-words that have a similar (or different) profile. Similarly, the standard coordinates of the six European countries can only be used to identify which countries have a similar (or different) profile.

There is a second problem with using the standard coordinates to graphically summarise the association between categorical variables. With a_{is}, say, being constrained by Eq. (2.8), we showed that the variance of the coordinates along each of the S dimensions are all identical and equal to one. Thus, each dimension in a plot constructed using standard coordinates of Table 2.3 reflects exactly $100/S\% = 100/5\% = 20\%$ of the similarities/differences in the (weighted centred) row profiles, and column profiles. Such a quantity is too low to be of any practical use. Nor is it possible to identify which dimension one should use to graphically summarise the association since they all equally account for the association. For example, the quality of a plot constructed using coordinates along the fourth and fifth dimensions will be equivalent to the quality of a plot constructed using the first and second dimensions, but the configuration of points could be very different. We can also make a similar statement regarding the column scores which are constrained by Eq. (2.9).

Therefore, while it may be appealing to use standard coordinates to visualise the association between categorical variables it is strongly advised that such plots not be constructed. We shall now describe the most common way of visualising the association between the row and column variables of a two-way contingency table; such a strategy takes into consideration the correlation between a_{is} and b_{js} along each of the S dimensions of the optimal correspondence plot.

2.5.2 Principal Coordinates

While standard coordinates should not be used to visually depict the association between the variables of a contingency table, such a visualisation can be simply obtained by multiplying them by the singular values. That is, the coordinate of the ith row category, and jth column category, on the sth dimension of the correspondence plot is defined by

$$f_{is} = a_{is}\lambda_s \tag{2.12}$$

$$g_{js} = b_{js}\lambda_s , \tag{2.13}$$

respectively; here λ_s is defined by Eq. (2.10). These coordinates are referred to as *principal coordinates* and are used to construct a low-dimensional correspondence plot consisting of no more than S dimensions. Of course, to make the visualisation of the association easy, and capture all of the association that exists in the contingency table, it would be ideal if $S = 2$ (or at most $S = 3$). However, this will not arise unless the variable with the smallest number of categories consists of three (or four) categories. In general the optimal correspondence plot will consists of $S > 3$ dimensions since, for many practical applications, either $I > 4$ and/or $J > 4$.

The principal coordinates defined by Eq. (2.12) and Eq. (2.13) are related to the Pearson residuals through

$$\gamma_{ij} - 1 = \sum_{s=1}^{S} a_{is} \lambda_s b_{js} = \sum_{s=1}^{S} \frac{f_{is} g_{js}}{\lambda_s} \tag{2.14}$$

and shows that when f_{is} and g_{js} are zero along each dimension of the optimal correspondence plot (consisting of exactly S dimensions), for all $i = 1, \ldots, I$ and $j = 1, \ldots, J$ then there is complete independence ($p_{ij} = p_{i\bullet} p_{\bullet j}$) in the (i, j)th cell of the contingency table. Therefore, the origin of the display coincides with the point where all of the principal coordinates would be located if there was complete independence between the two variables of the contingency table. From this property we can see that points close to the origin play a relatively small role in defining the association structure between the variables than points that are positioned away from the origin. In fact, we can also show that the principal coordinates are centred at the origin of the optimal correspondence plot. To do so, consider the row principal coordinates. Then the expectation of f_{is} is

$$E(f_{is}) = \sum_{i=1}^{I} \sum_{s=1}^{S} p_{i\bullet} f_{is}$$

$$= \sum_{i=1}^{I} \sum_{s=1}^{S} p_{i\bullet} a_{is} \lambda_s$$

$$= \sum_{s=1}^{S} \lambda_s \left(\sum_{i=1}^{I} p_{i\bullet} a_{is} \right)$$

$$= 0 .$$

We can also show that the variance of the ith row principal coordinate in this plot is equivalent to the total inertia of the contingency table since

$$\mathrm{Var}(f_{is}) = E(f_{is}^2) - [E(f_{is})]^2$$

$$= \sum_{i=1}^{I} \sum_{s=1}^{S} p_{i\bullet} f_{is}^2 - 0$$

$$= \sum_{i=1}^{I} \sum_{s=1}^{S} p_{i\bullet} a_{is}^2 \lambda_s^2$$

$$= \sum_{s=1}^{S} \lambda_s^2 \left(\sum_{i=1}^{I} p_{i\bullet} a_{is}^2 \right)$$

$$= \sum_{s=1}^{S} \lambda_s^2$$

$$= \frac{X^2}{n}$$

using Eq. (2.11). Similar derivations can also be made showing that the column principal coordinates are centred around the origin of the correspondence plot and their variation can be measured using the total inertia, X^2/n. The term λ_s^2 is the amount of the total inertia that is explained by the sth dimension. Therefore, we refer to it as the sth *explained inertia*;

it is also commonly referred to as the *s*th *principal inertia*. To help better understand the contribution that this explained inertia makes to the total inertia it is often expressed as a percentage so that

$$100 \times \frac{\lambda_s^2}{X^2/n}$$

reflects the percentage contribution that the *s*th dimension makes to the association between the variables. For a two-dimensional correspondence plot, say, we can therefore determine how much of the association (as a percentage) it visually summarises using the quantity

$$100 \times \frac{\lambda_1^2 + \lambda_2^2}{X^2/n} \ .$$

Since $\lambda_1 > \lambda_2$ are the two largest singular values this means that a two-dimensional correspondence plot will give the best possible two-dimensional visual summary of the association.

There are other important features that make using principal coordinates ideal for the visualisation of the association. One such feature is that their distance from each other, and from the origin are interpretable. Suppose we consider the row principal coordinates defined by Eq. (2.12). The squared Euclidean distance between the point for the *i*th row and the *i'*th row in the optimal correspondence plot is

$$d_I^2(i, \ i') = \sum_{s=1}^{S} (f_{is} - f_{i's})^2 \tag{2.15}$$

while this distance is directly related to the difference between two (centred) row profiles defined by Eq. (2.1). That is

$$d_I^2(i, \ i') = \sum_{j=1}^{J} \frac{1}{p_{\bullet j}} \left(\frac{p_{ij}}{p_{i\bullet}} - \frac{p_{i'j}}{p_{i'\bullet}} \right)^2$$

$$= \sum_{j=1}^{J} \frac{1}{p_{\bullet j}} \left[\left(\frac{p_{ij}}{p_{i\bullet}} - p_{\bullet j} \right) - \left(\frac{p_{i'j}}{p_{i'\bullet}} - p_{\bullet j} \right) \right]^2 \ . \tag{2.16}$$

These two distance formulae show that if two points are close to each other in an optimal correspondence plot then their profiles will be similar. Similarly, if two points are at a distance from one another in such a plot then these formulae imply that their profile is very different from each other. Such an interpretation is referred to as the *property of distributional equivalence* in the correspondence analysis literature; see, for example, Lebart et al. (1984, p. 35), Greenacre (1984, p. 95) and Beh and Lombardo (2014, Section 4.6) for a description of this property. Note that such an interpretation is made in terms of the optimal correspondence plot, not a two-dimensional correspondence plot. In many practical applications of correspondence analysis a two-dimensional correspondence plot will *approximately* reflect this property, but not always exactly. As we shall show in our example in Section 2.6, it is important to keep in mind not just the configuration shown in a two-dimensional correspondence plot but also the quality of the plot. Unfortunately, in many applications of correspondence analysis, this feature is rarely examined. The omission of the quality of visualisation given by the two-(or three-) dimensional correspondence plot is akin to neglecting the magnitude of the R^2 value when assessing the quality of a regression model.

We have briefly described above that the origin is an important coordinate in the correspondence plot. So, to further help with the interpretation of the origin, we shall now consider the distance of the ith row principal coordinate from the origin in an S-dimensional correspondence plot. In this case Eq. (2.15) becomes

$$d_I^2(i, \ 0) = \sum_{s=1}^{S} (f_{is} - 0)^2$$

$$= \sum_{s=1}^{S} f_{is}^2$$

while Eq. (2.16) simplifies to

$$d_I^2(i, \ 0) = \sum_{j=1}^{J} \frac{1}{p_{\bullet j}} \left(\frac{p_{ij}}{p_{i\bullet}} - p_{\bullet j} \right)^2$$

$$= \sum_{j=1}^{J} p_{\bullet j} \left(\frac{p_{ij}}{p_{i\bullet} p_{\bullet j}} - 1 \right)^2 .$$

Note that by considering these equations in the context of Pearson's chi-squared statistic – see Eq. (2.6) – the total inertia can be expressed as the weighted sum-of-squares of the row principal coordinates. It can also be expressed as the weighted sum of the squared Euclidean distance of each row (or column) principal coordinate from the origin. That is,

$$\frac{X^2}{n} = \sum_{i=1}^{I} \sum_{s=1}^{S} p_{i\bullet} f_{is}^2$$

$$= \sum_{i=1}^{I} p_{i\bullet} d_I^2(i, \ 0) .$$

So, the origin is the point where all principal coordinates will lie if there is complete independence between the two categorical variables. If the points are far from the origin then they show a clear deviation from what we would expect if both categorical variables were completely independent. Points that are at, or near, the origin show that the frequencies in a row of the contingency table are exactly, or consistent with, what is expected under complete independence. A similar argument, and distance expressions, can also be made for the column categories of a contingency table. We may also derive such results by noting that if all $\lambda_s = 0$, for $s = 1, \ 2, \ \ldots \ , \ S$, then all principal coordinates will be zero (hence lying at the origin) and the total inertia defined will be zero; see Eq. (2.11).

2.6 Practicalities of the Low-Dimensional Plot

2.6.1 The Two-Dimensional Correspondence Plot

Suppose we consider again the traditional European food data of Table 1.1. The $I = 28$ row principal coordinates – defined using Eq. (2.12) – and the $J = 6$ column principal coordinates – defined using Eq. (2.13) – are summarised in the rows of Table 2.5 and

Table 2.5 The principal coordinates, f_{is}, of Table 1.1.

Code	Free-word	Dim 1	Dim 2	Dim 3	Dim 4	Dim 5
1	Ancient	0.657	−0.953	0.721	−0.606	−0.205
2	Christmas	−0.476	−0.698	−0.632	0.311	−0.070
3	Cooking	0.006	0.526	−0.117	−0.822	−0.119
4	Country	−0.551	−0.421	−0.546	0.217	−0.083
5	Culture	0.731	0.114	0.063	−0.228	−0.266
6	Dinner	−1.070	0.166	0.262	0.721	−0.530
7	Dish	−0.739	0.860	0.043	−0.373	−0.235
8	Family	−0.235	0.058	−0.029	−0.189	0.104
9	Feast	0.596	0.955	−0.682	−0.107	0.264
10	Good	0.136	−0.588	0.323	−0.172	0.489
11	Grandmother	0.611	0.049	−0.237	0.189	0.027
12	Habit	0.954	−0.213	0.019	−0.025	−0.281
13	Healthy	−0.102	0.037	0.407	0.358	−0.017
14	Holidays	0.095	0.647	0.271	0.927	−0.412
15	Home	0.267	−0.394	0.480	−0.262	−0.601
16	Home-made	−0.318	−0.310	0.818	0.166	0.407
17	Kitchen	0.263	0.169	0.336	0.624	0.256
18	Meal	−0.532	0.549	0.298	−0.060	−0.197
19	Natural	−0.048	−0.125	0.487	0.087	0.546
20	Old	0.038	−0.248	0.062	−0.282	−0.352
21	Old-fashioned	1.043	0.337	−0.867	1.134	0.251
22	Quality	0.583	0.269	−0.291	−0.371	−0.042
23	Recipe	−0.684	0.911	−0.043	−0.520	−0.167
24	Regional	−0.159	0.335	0.016	−0.495	0.961
25	Restaurant	0.412	0.574	−0.388	0.026	0.347
26	Rural	−0.276	−0.362	−0.635	−0.246	0.117
27	Simple	−0.041	0.123	0.471	0.349	0.743
28	Tasty	−0.001	0.272	0.470	0.353	−0.070

Table 2.6, respectively. These coordinates were calculated in R using the CAvariants package.

The two-dimensional correspondence plot of Table 1.1 is given by Figure 2.6 and from it we can assess how particular rows and columns are associated with one another (through the proximity of their points). For example, the plot suggests that those who participated in the study in *Norway* associate traditional European food with the free-words *Christmas* (Code 2), *Country* (Code 4), *Home-made* (Code 16) and *Rural* (Code 26) since these row and column pairs are close to one another in Figure 2.6. Similarly, those in *France* perceived

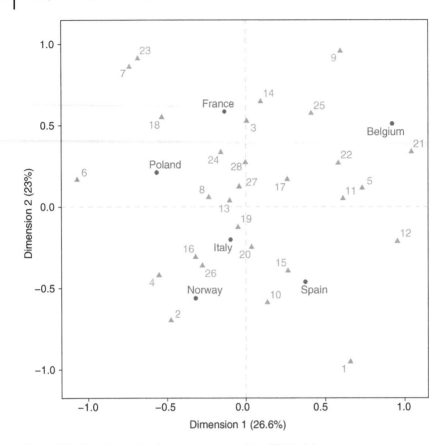

Figure 2.6 Two-dimensional correspondence plot of Table 1.1.

traditional European food with the free-words *Cooking* (Code 3), *Regional* (Code 24) and *Tasty* (Code 28) while *Kitchen* (Code 17) and *Quality* (Code 22) were strongly linked to the perception of traditional European food in *Belgium*. The free-word that the residents of *Italy* most strongly associated with traditional European food was found to be *Natural* (Code 19) while those in *Spain* perceived such food with *Home* (Code 15). While their relative distances are far from each other, *Dinner* (Code 6) is best associated with *Poland*, *Old-fashioned* is best associated with *Belgium* while *Dish* (Code 7), *Recipe* (Code 23) and *Meal* (Code 18) are most strongly linked to *France*.

There are also free-words that can be linked to more than one country. For example, *Simple* (Code 27) is associated with how *France* and *Poland* perceive traditional European food. Similarly, the free-word *Old* (Code 20) are strongly linked with *Norway*, *Spain* and *Italy* while *Healthy* is equally associated with *Poland* and *Italy*.

Figure 2.6 shows that the way in which *Italy* and *Spain* perceive traditional European food is very similar. Such a conclusion is based on their relative closeness to each other in the two-dimensional correspondence plot. This plot also suggests that the free-words *Cooking* (Code 3) and *Holidays* (Code 14) are very similar in their profile due to the close proximity that their points share. However, for both cases, we discussed in Section 2.2.3 that their profiles appeared to be very different. So why the apparent contradiction? Firstly, the

Table 2.6 The principal coordinates, g_{js}, of Table 1.1.

Country	Dim 1	Dim 2	Dim 3	Dim 4	Dim 5
Belgium	0.921	0.510	−0.176	0.568	0.041
France	−0.135	0.585	−0.293	−0.650	0.106
Italy	−0.096	−0.201	0.833	0.069	1.063
Norway	−0.318	−0.561	−0.768	0.242	0.159
Poland	−0.565	0.212	0.302	0.303	−0.239
Spain	0.375	−0.462	0.241	−0.287	−0.216

difference between the profile comparison and our interpretation of Figure 2.6 has nothing to do with the decomposition of the Pearson residuals. Instead the source of the apparent confusion can be identified by considering the quality of the two dimensional correspondence plot of Figure 2.6. Since, for the traditional European food data, $S = \min(28, 6) − 1 = 5$, the optimal correspondence plot consists of five dimensions, not two dimensions as are displayed in Figure 2.6. Therefore, the two-dimensional plot does not reflect 100% of the association that exists between the variables. Quantifying the magnitude of this association is achieved by calculating the *total inertia*,

$$\frac{X^2}{n} = \frac{663.759}{743} = 0.893 .$$

The total inertia, like the chi-squared statistic, quantifies the strength of the association between the variables. However, unlike X^2 which is bounded by $0 \leq X^2 \leq nS$, the total inertia takes on a value between 0 and S (inclusive) for the analysis of a two-way contingency table. With $S = 5$, the total inertia appears to be quite small and suggests that there may not be a "strong" association between *Free-word* and *Country*. However, their does exist a statistically significant association – a conclusion we found by performing a chi-squared test of independence between the two variables. The proportion of the total inertia that is described by Figure 2.6, which consists of the first two sets of principal coordinates summarised in Table 2.5 and Table 2.6, is calculated by finding the percentage contribution of the first two squared singular values to the total inertia. So, since the first two explained inertia values are $\lambda_1^2 = 0.237$ and $\lambda_2^2 = 0.205$, this percentage contribution is

$$100 \times \frac{0.237 + 0.205}{0.893} = 49.50\% .$$

Thus, the two-dimensional display of Figure 2.6 visually describes about half of the association that exists between the two variables of Table 1.1. Such a quantity therefore shows that the third and higher dimensions, collectively, account for the remaining 50.50% of the total inertia that is not accounted for in Figure 2.6. This is a topic we shall talk more about shortly but it is an extremely importance facet to correspondence analysis, and one that is often ignored in practice. However, we can identify the proportion of the total inertia that each dimension of Figure 2.6 makes. Since the first singular value is the largest of all the singular values, the first dimension of Figure 2.6 reflects more of the association than any other dimension. This first dimension visually summarises

$$100 \times \frac{0.237}{0.893} = 26.55\%$$

of the association between the *Free-word* and *Country* variables. Similarly, the second dimension accounts for

$$100 \times \frac{0.205}{0.893} = 22.95\%$$

of the association. These two percentages are also summarised along each of the two dimensions of Figure 2.6.

2.6.2 What is NOT Being Shown in a Two-Dimensional Correspondence Plot?

With only about 50% of the association between the variables of Table 1.1 being depicted in Figure 2.6, this correspondence plot is not considered to be a good quality two-dimensional display. Despite this, one of the biggest problems often observed in the practical application of correspondence analysis is that the quality of the display is not included as part of the findings of the analysis; we have made this point a few times already. Such omissions may be unintentional since the researcher may not be aware that the quality of a low-dimensional plot is an important feature of the analysis. Perhaps such detail is intentionally avoided since the plot provides a poor display of the association and the researcher does not wish to draw attention to it. In either case, we can quantify where the two-dimensional plot fails to capture the association structure by identifying those row and column categories that are poorly represented. We can also improve the quality of the two-dimensional correspondence plot by simply including a third dimension to the display – we shall discuss this option in Section 2.6.3.

To assess where the two-dimensional correspondence analysis fails to adequately represent the association between the *Free-word* and *Country* variables of Table 1.1, we will determine how much of the association is contained in the third and higher dimensions. There are several ways that this can be achieved but we shall consider here the reconstruction, or reconstitution, of the elements of $\gamma_{ij} - 1$ given the information contained in the first two dimensions of the optimal correspondence plot (Figure 2.6). From Eq. (2.14), this reconstitution gives the approximation

$$\hat{\gamma}_{ij} - 1 = \sum_{s=1}^{2} a_{is} \lambda_s b_{js} = \frac{f_{i1} g_{j1}}{\lambda_1} + \frac{f_{i2} g_{j2}}{\lambda_2} \tag{2.17}$$

where $\hat{\gamma}_{ij} - 1$ is the approximation of the (i, j)th Pearson residual given only the information in the first two dimensions. Therefore, if we calculate

$$\delta_{ij} = \sqrt{p_{i\bullet} p_{\bullet j}} ((\gamma_{ij} - 1) - (\hat{\gamma}_{ij} - 1))$$

$$= \sqrt{p_{i\bullet} p_{\bullet j}} \left(\gamma_{ij} - \left(1 + \frac{f_{i1} g_{j1}}{\lambda_1} + \frac{f_{i2} g_{j2}}{\lambda_2} \right) \right)$$

then we can get an idea of where Figure 2.6 fails to capture the association. Note that, from Eq. (2.11), the sum-of-squares of the δ_{ij} is

$$\frac{X_{(3:S)}^2}{n} = \sum_{i=1}^{I} \sum_{j=1}^{J} \delta_{ij}^2 = \sum_{s=3}^{S} \lambda_s^2 \tag{2.18}$$

and is that part of the total inertia not captured in the first two dimensions. Note that the sum-of-squares of the δ_{ij} values in Table 2.7, using Eq. (2.18), is 0.451. This value is that part of the total inertia that is explained by the third and higher dimensions. Therefore

$$100 \times \frac{0.451}{0.893} = 50.50\%$$

of the total inertia is accounted for by the third, fourth and fifth dimensions of the optimal correspondence plot of Table 1.1. Thus, 50.50% of the total inertia is not accounted for in the two-dimensional correspondence plot of Figure 2.6. This should be of no surprise since

Table 2.7 The δ_{ij} values of Table 1.1.

		Country					
Code	Free-word	Belgium	France	Italy	Norway	Poland	Spain
1	Ancient	−0.063	0.026	0.023	−0.102	0.015	0.074
2	Christmas	0.058	−0.009	−0.084	**0.118**	−0.019	−0.064
3	Cooking	−0.083	**0.118**	−0.040	−0.028	−0.063	0.064
4	Country	0.042	0.000	−0.074	0.094	−0.018	−0.047
5	Culture	−0.019	0.014	−0.028	−0.021	0.005	0.027
6	Dinner	0.045	−0.094	−0.043	−0.018	0.083	0.000
7	Dish	−0.034	0.034	−0.032	−0.027	−0.006	0.038
8	Family	−0.025	0.043	0.019	−0.001	−0.034	0.007
9	Feast	0.009	0.048	−0.018	0.079	−0.059	−0.039
10	Good	−0.021	0.017	**0.108**	−0.032	−0.026	−0.003
11	Grandmother	0.026	−0.011	−0.017	0.043	−0.004	−0.029
12	Habit	−0.007	−0.007	−0.056	−0.018	0.025	0.027
13	Healthy	0.023	−0.067	0.038	−0.038	0.053	−0.001
14	Holidays	0.068	**−0.125**	−0.029	−0.009	0.098	−0.019
15	Home	−0.036	−0.007	−0.045	−0.081	0.046	0.070
16	Home-made	−0.004	−0.058	**0.151**	−0.091	0.039	0.007
17	Kitchen	0.036	−0.062	0.053	−0.005	0.033	−0.029
18	Meal	−0.014	−0.012	−0.004	−0.043	0.026	0.029
19	Natural	0.000	−0.020	**0.116**	−0.035	0.000	−0.014
20	Old	−0.042	0.031	−0.070	−0.045	0.012	0.061
21	Old-fashioned	0.099	−0.067	−0.023	**0.128**	0.004	**−0.105**
22	Quality	−0.018	0.041	−0.024	0.014	−0.028	0.008
23	Recipe	−0.034	0.044	−0.024	−0.016	−0.019	0.031
24	Regional	−0.030	0.067	**0.115**	0.009	−0.078	−0.022
25	Restaurant	0.012	0.020	0.015	0.048	−0.036	−0.034
26	Rural	−0.006	0.091	−0.056	0.102	−0.089	−0.033
27	Simple	0.017	−0.032	0.104	−0.012	0.002	−0.030
28	Tasty	0.024	−0.087	0.042	−0.059	0.073	0.007

we showed that the first two dimensions of the optimal correspondence plot – given by Figure 2.6 – visually summarises 49.50% of the total inertia in the contingency table. We could have also found this part of the total inertia by calculating the sum of λ_3^2, λ_4^2 and λ_5^2. That is

$$\lambda_3^2 + \lambda_4^2 + \lambda_5^2 = 0.185 + 0.164 + 0.101 = 0.450$$

which (other than the presence of a rounding error at the third decimal place) is exactly what we got from Eq. (2.18).

Table 2.7 shows that the value of δ_{ij} for each *Free-word–Country* pair. Those values in bold highlight the nine poorest fitting row/column associations captured in Figure 2.6. For example, with a $\delta_{16,\ Italy}$ value of 0.151, the association between *Italy* and *Home-made* (Coded 16) is poorly depicted. So, while Figure 2.6 shows that, due to their relatively close proximity, these two categories are strongly associated, the association is not as close as we would expect when we take into account their position in the third and higher dimensions. Table 2.7 also suggests that *Italy* is the poorest represented country in Figure 2.6 since it consists of more of the most extreme values of δ_{ij} than any other country.

Table 2.8 summarises the contribution, and percentage contribution, of each of the six countries to the total inertia given the information they contain in the third and higher dimensions. These contributions are quantified by

$$\delta_{J(j)}^2 = \sum_{i=1}^{I} \delta_{ij}^2$$

so that

$$\frac{X_{(3:S)}^2}{n} = \sum_{j=1}^{J} \delta_{J(j)}^2 .$$

These values show, for example, that *Italy* is the poorest fitting country in Figure 2.6 contributing to a quarter of the remaining contribution not captured in the two-dimensional plot. One may also note that, as we described in the previous paragraph, *Italy* also contains four of the nine bolded figures in Table 2.7. On the other hand, *Belgium* is the best represented in the first two-dimensions since it only contributes to about 10% of the association contained in the third and higher dimensions.

Table 2.9 summarises the free-words that are well fitted, and poorly fitted, in Figure 2.6. Such a fit is assessed by

$$\delta_{I(i)}^2 = \sum_{j=1}^{J} \delta_{ij}^2$$

Table 2.8 Contribution ($\delta_{J(j)}^2$) and percentage contribution, (%) of each *Country* category to the third and higher dimensions of the optimal correspondence plot.

	Country						
Quantity	**Belgium**	**France**	**Italy**	**Norway**	**Poland**	**Spain**	**Total**
$\delta_{J(j)}^2$	0.045	0.088	0.113	0.100	0.057	0.048	0.451
%	9.978	19.512	25.055	22.173	12.639	10.643	100.00

Table 2.9 Contribution ($\delta^2_{I(i)}$) and percentage contribution (%) of each *Free-word* category to the third and higher dimensions of the optimal correspondence plot.

	Free-word (Code)			
Quantity	**Ancient (1)**	**Christmas (2)**	**Cooking (3)**	**Country (4)**
$\delta^2_{I(i)}$	0.021	0.029	0.031	0.018
%	4.656	6.430	6.874	3.991
	Culture (5)	Dinner (6)	Dish (7)	Family (8)
$\delta^2_{I(i)}$	0.003	0.020	0.006	0.004
%	0.665	4.435	1.330	0.887
	Feast (9)	Good (10)	Grandm. (11)	Habit (12)
$\delta^2_{I(i)}$	0.017	0.014	0.004	0.005
%	3.104	3.104	0.887	1.109
	Healthy (13)	Holidays (14)	Home (15)	Home-made (16)
$\delta^2_{I(i)}$	0.011	0.031	0.017	0.036
%	2.439	6.874	3.769	7.982
	Kitchen (17)	Meal (18)	Natural (19)	Old (20)
$\delta^2_{I(i)}$	0.010	0.004	0.015	0.014
%	2.217	0.887	3.326	3.104
	Old-fash. (21)	Quality (22)	Recipe (23)	Regional (24)
$\delta^2_{I(i)}$	0.042	0.004	0.005	0.025
%	9.313	0.887	1.109	5.543
	Restaurant (25)	Rural (26)	Simple (27)	Tasty (28)
$\delta^2_{I(i)}$	0.005	0.031	0.013	0.019
%	1.109	6.874	2.882	4.213

so that

$$\frac{X^2_{(3:S)}}{n} = \sum_{i=1}^{I} \delta^2_{I(i)} \, .$$

These values highlight those free-words that are relatively poorly fitted are, in order of "poor-fit", *Old-fashioned* (Code 21), *Home-made* (Code 16), *Cooking* (Code 3) and *Holidays* (Code 14). These four free-words combined account for nearly a third (31.043%) of the inertia of all of the words in the third and higher dimensions of the optimal correspondence plot. Those free-words in Table 1.1 that are well visualised in Figure 2.6, in order of "best-fit", are *Culture* (Code 5), *Family* (Code 8), *Grandmother* (Code 11) *Meal* (Code 18) and *Quality* (Code 22). These free-words each contribute to less than 1% of the fit of the words in the third and higher dimensions. These findings correspond well to our comments above.

2.6.3 The Three-Dimensional Correspondence Plot

When a two-dimensional correspondence plot is of poor quality, it can often be easily improved by adding a third dimension to the display. Figure 2.7 provides a

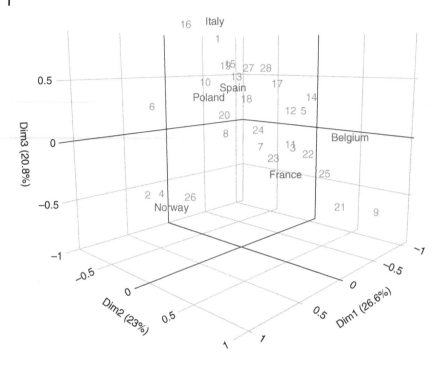

Figure 2.7 Three-dimensional correspondence plot of Table 1.1.

three-dimensional correspondence plot that visually depicts the association between the *Country* and *Free-word* variables of Table 1.1. Since $\lambda_3^2 = 0.185$, Figure 2.7 reflects

$$100 \times \frac{\lambda_1^2 + \lambda_2^2 + \lambda_3^2}{\sum\limits_{s=1}^{S} \lambda_s^2} = 100 \times \frac{0.237 + 0.205 + 0.185}{0.893}$$

$$= 70.213\%$$

of the total inertia of Table 1.1.

An obvious question then is "what threshold should be used to assess whether a two- or three-dimensional correspondence plot is of good quality (or not)?". There are many formal procedures available to help to make such a decision. For example, one may refer to Lebart (1976), Lebart et al. (1984), Blasius (1994, pp. 28–29), Ciampi et al. (2005), Lorenza-Seva (2011) and Beh and Lombardo (2014, pp. 145–147) for various technical discussions on this issue. Benzécri (1992, p. 398) argued that the decision should be made based on the researchers personal judgement rather than by any mathematical procedure. We would generally advocate such an approach, especially for those who are aware of the nuances of correspondence analysis. For those who are not, one must be aware that leaving the decision based purely on a subjective rationale may result in important, or valuable, information at higher dimensions being neglected because of the failure to investigate these higher dimensions. As we have stated above, this commonly arises in many practical applications of correspondence analysis. Our recommendation in this book is that a visual summary that displays at least 70% of the association (as quantified by the total inertia) will be deemed a good quality visual representation of this association.

2.7 The Biplot Display

2.7.1 Definition

In correspondence analysis, using principal coordinates is the most common way to visualise the association between the row and column variables of a two-way contingency table. However, such coordinates do not provide for a clear interpretation to be made between a row point and a column point. Given that a correspondence plot is designed to provide a joint visualisation of the position of the row and column profiles in a low-dimensional space, it makes sense that if we can interpret the distance between two row (or column) points, then we should also be able to interpret the distance between a row point and a column point.

To provide a meaningful interpretation of the distance between a row point and a column point in a low-dimensional correspondence plot we can consider alternative coordinates to those given by Eq. (2.12) and Eq. (2.13). A simple alternative is to simply rescale the sth singular value, λ_s, such that the ith row and jth column coordinate along the sth dimension is defined by

$$\tilde{f}_{is} = a_{is}\lambda_s^\epsilon \tag{2.19}$$

$$\tilde{g}_{js} = b_{js}\lambda_s^{1-\epsilon}, \tag{2.20}$$

respectively, for $0 \leq \epsilon \leq 1$. These coordinates are referred to as *biplot coordinates* and when using them to define the position of the row and column categories, the resulting low-dimensional plot is referred to as a *biplot*. The biplot has received a lot of (positive) attention in the correspondence analysis literature for quite a few years now and one may refer to, for example, Gabriel (1971) (who first proposed their use), Greenacre (2010), Gower et al. (2011) and Gower et al. (2016) for more information on their construction and use. By defining the coordinates in this manner the inner product of $\tilde{f}_{is}$ and $\tilde{g}_{js}$ is

$$\sum_{s=1}^{S}\tilde{f}_{is}\tilde{g}_{js} = \sum_{s=1}^{S}a_{is}\lambda_s b_{js} = \gamma_{ij} - 1$$

and so reflects the strength of the interaction between the ith row and jth column categories.

The choice of ϵ determines the emphasis that is placed on a variable. The two most widely used values of ϵ are

- $\epsilon = 0$ and defines the coordinate of the ith row and jth column along the sth dimension by $\tilde{f}_{is} = a_{is}$ and $\tilde{g}_{js} = b_{js}\lambda_s$. The plot using this value of ϵ is called a *column isometric biplot*. Such a biplot is therefore constructed using the standard coordinates of the row categories and the principal coordinates of the column categories; each row category is depicted as a projection from the origin to their position in the biplot defined by their standard coordinate while each column category is depicted as a single point at their principal coordinate. For such a specification, the (i, j)th Pearson residual is expressed as

$$\gamma_{ij} - 1 = \sum_{s=1}^{S}a_{is}\tilde{g}_{js}$$

and is the inner product of the row standard coordinate and the column principal coordinate in the optimal biplot. Therefore, if this inner product is large (so that the position of the jth column principal coordinate is in close proximity to the projection of the ith row standard coordinate) then there is a close interaction between this row/column pair.

- $\epsilon = 1$ and defines the coordinate of the ith row and jth column along the sth dimension by $\tilde{f}_{is} = a_{is}\lambda_s$ and $\tilde{g}_{js} = b_{js}$ yielding a *row isometric biplot*. Such a biplot is therefore constructed using the standard coordinates of the column categories and the principal coordinates of the row categories; each column category is depicted as a projection from the origin to their position in the biplot defined by their standard coordinate while each row category is depicted as a single point at their principal coordinate. For this value of ϵ, the (i, j)th Pearson residual can be expressed by

$$\gamma_{ij} - 1 = \sum_{s=1}^{S} \tilde{f}_{is} b_{js}$$

and is the inner product of the column standard coordinate and the row principal coordinate in the optimal biplot. Therefore, if this inner product is large (such that the position of the ith row principal coordinate is in close proximity to the projection of the jth column standard coordinate) then there is a close association between this row/column pair.

To provide more insight into the interpretation of the distance of a standard coordinate from the projection of a principal coordinate from the origin, we now examine the angle between this projection and the position of the standard coordinate. We do so by expressing these coordinates as elements of a vector. Suppose we define $\tilde{\mathbf{f}}_i = (\tilde{f}_{i1}, \tilde{f}_{i2}, \ldots, \tilde{f}_{iS})$ to be the vector of the ith row principal coordinate in the optimal correspondence plot while $\mathbf{b}_j = (b_{j1}, b_{j2}, \ldots, b_{jS})$ is the vector of the jth column standard coordinate in this space. Therefore, the angle, θ_{ij} between the projection of $\tilde{\mathbf{f}}_i$ and the position of $\mathbf{b}_j$ is

$$\cos\theta_{ij} = \frac{\tilde{\mathbf{f}}_i \mathbf{b}_j}{||\tilde{\mathbf{f}}_i||^2 \bullet ||\mathbf{b}_j||^2}$$

where

$$||\mathbf{b}_j||^2 = b_{j1}^2 + \ldots + b_{jS}^2$$

and

$$||\tilde{\mathbf{f}}_i||^2 = f_{i1}^2 + \ldots + f_{iS}^2 \ .$$

Therefore, geometrically, the cosine of the angle between the ith row principal coordinate and the jth standard coordinate is equivalent to the correlation between them.

2.7.2 Isometric Biplots of the Traditional European Food Data

Figure 2.8 and Figure 2.9 give, respectively, the row and column isometric biplots of Table 1.1. The advantage of these plots is that the proximity between a point representing a row category and a column category can be made by observing the shortest distance of a point (principal coordinate) from the projection (line) of a coordinate from the origin. Both biplots show that respondents in *Norway* feel that traditional European food is associated with the free-words *Christmas* (Code 2), *Country* (Code 4), *Home-made* (Code 16) and *Rural* (Code 26). This is apparent by observing, say, the configuration

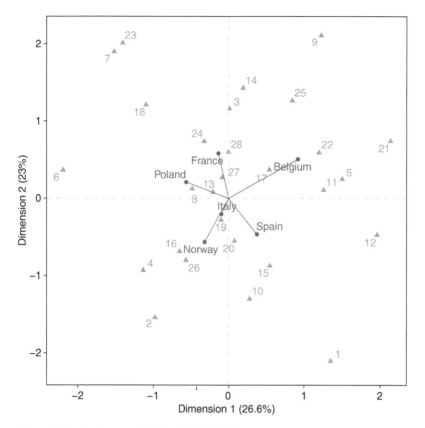

Figure 2.8 Row isometric biplot of of Table 1.1.

of points of the row isometric biplot of Figure 2.8. It shows that the position of each of these *Free-word* categories lies closer to the projection of the standard coordinate from the origin of *Norway* than does any other *Free-word* category. Similarly, there is an interaction between how respondents from *Belgium* perceive traditional European food with the free-words *Old-fashioned* (Code 21) and *Quality* (Code 22).

An important aspect of the biplot is that it may be tempting for one to interpret a large distance between a row principal coordinate from the location of a column standard coordinate as always meaning that there is a lack of interaction between a *Free-word/Country* pair of categories. For example, the free-word *Rural* (Code 26), in the bottom-left quadrant of Figure 2.8, is located some distance away from where *Spain* is located in Figure 2.8 and so one may interpret that such a large distance means that there is a weak interaction between these two categories. However, since Figure 2.8 is a biplot, it is the distance of the points from the *Country* projections that is the important factor in helping to identify the nature of the interaction between these categories. We could certainly project both *Italy*, and *Norway*, further from the origin than we have done and their interaction with the free-words would remain unchanged. That is, there would still be a strong interaction between *Rural* (Code 26), say, and *Italy*.

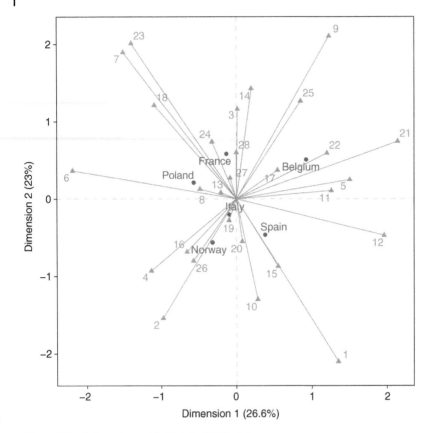

Figure 2.9 Column isometric biplot of Table 1.1.

Figure 2.8 also shows that those from *France* perceive traditional European food with the free-words *Cooking* (Code 3), *Holidays* (Code 14), *Regional* (Code 24), *Simple* (Code 27) and *Tasty* (Code 28). There are some free-words that do not have a strong interaction with any particular country but are considered as having a near "average" profile. For example, the free-words *Family* (Code 8), *Healthy* (Code 13), *Natural* (Code 19) and *Simple* (Code 27) all lie very close to the origin. While some appear to have a weak interaction with particular countries, their contribution to the association structure between the *Free-word* and *Country* variables appears to be minimal. Sections 2.9 and 2.10 discuss the interpretation of what it means for a principal coordinate to be "close to the origin".

Note that the conclusions we have made from interpreting the row isometric biplot of Figure 2.8 also apply to the interpretation of the column isometric biplot of Figure 2.9. For example, it shows that the position of the column principal coordinates of *Norway* and *Italy* lies very close to the projection of *Rural* (Code 26) from the origin. Similarly the position of *France* is located in close proximity to the projection of the free-words *Cooking* (Code 3), *Holidays* (Code 14), *Regional* (Code 24) and *Simple* (Code 27) from the origin.

2.7.3 What is NOT Being Shown in a Two-Dimensional Biplot?

In Section 2.6.2 we described how to identify those row and column categories that are not adequately visualised in a two-dimensional correspondence plot. Such a procedure can be extended for a three-dimensional correspondence plot; see Section 2.6.3. For an isometric biplot, we can also quantify how well the two, or three-dimensional, display summarises the association by examining those categories that are well, or poorly, represented in the biplot. For a two-dimensional isometric biplot, examining how much information is lost in the third and higher dimensions can be achieved by first approximating Pearson's residual for the (i, j)th cell by

$$\gamma_{ij} - 1 \approx \sum_{s=1}^{2} a_{is}\lambda_s b_{js}$$
$$= \tilde{f}_{i1}\tilde{g}_{j1} + \tilde{f}_{i2}\tilde{g}_{j2} \ .$$

This summation is the inner product of the ith row biplot coordinate and the jth column biplot coordinate along the first two dimensions. For the column isometric biplot of Figure 2.9 such an approximation can be made using the first two row standard coordinates and the first two column principal coordinates such that

$$\gamma_{ij} - 1 \approx a_{i1}g_{j1} + a_{i2}g_{j2}$$

while for a row isometric plot, the approximation is

$$\gamma_{ij} - 1 \approx f_{i1}b_{j1} + f_{i2}b_{j2} \ .$$

Note that these two equations are akin to Eq. (2.17). Quantifying the association that is missing from a two-dimensional isometric biplot can be done by keeping in mind the comments made above for the principal coordinates and these isometric approximations of the Pearson residual. Therefore, to avoid repeating the analysis, we shall not discuss this aspect any further for the biplot.

2.8 The Case for No Visual Display

When discussing the interpretation of the configuration of points in a correspondence plot, we focused our discussion on the meaning of the distance between points that come from the same variable. This is because there is a strong and accepted theoretical grounding for such an interpretation. However, there has been considerable debate on the issue of quantifying and interpreting the distance between a row point and a column point in a low-dimensional display. Much of this contention stems from a series of papers in the 1980's where Carroll et al. (1986) proposed, and further clarified in 1987, a set of row and column principal coordinates that would allow for the distance between a row and a column point in a correspondence plot to be interpretable. Their proposed coordinates stem from, in part, the theory underlying the structure of principal coordinates in multiple correspondence analysis; a topic that we will be discussing in Chapter 6. It was on these grounds that Greenacre (1989, 1990) strongly argued against their proposal. Despite their disagreement, Hoffman et al. (1995) point out that the difference of opinion is more philosophical than technical.

The argument for whether the distance between a row principal coordinate and a column principal coordinate can be measured and interpreted is not confined to this debate. There have also been extensive arguments made for no visual representation at all! Much of the literature on this topic is dominated by Shizuhiko Nishisato; see, for example, Nishisato (1988, 1995) and Nishisato (1994, Chapter 14) who discussed a variety of issues with the joint representation of row and column points that come from different clouds-of-points. More recently, Nishisato and Clavel (2003, 2010) also examined this issue and pointed out that part of the problem with a joint visual representation is concerned with the angle that separates them. They point out that $\cos^{-1}(\lambda_s)$ is the *angle of discrepancy between the row space and column space* along the sth dimension. For example, if we consider our analysis of Table 1.1, since $\lambda_1^2 = 0.237$ then, the angle of discrepancy along the first dimension of the row and column spaces is $\cos^{-1}(\sqrt{0.237}) = 60.87$ degrees which is quite large and so, even along the first dimension, there is not a strong case for making a meaningful or practical interpretation of the distance between a row and column principal coordinate.

Despite such arguments we shall continue to visualise the association between the row and column variables using a correspondence plot or biplot. In doing so, we shall not be making any further comment on the distance, or angle, between row-column points or spaces.

2.9 Detecting Statistically Significant Points

2.9.1 Confidence Circles and Ellipses

To assess how "close" to the origin a row or column point is in a correspondence plot, various approaches have been proposed. These approaches are, therefore, designed to detect those rows and columns that provide a statistically significant contribution to the association between the variables of a two-way contingency table. Lebart et al. (1984, pp. 182 – 186) proposed a very simple means of constructing $100(1 - \alpha)\%$ confidence circles for each of the categories. They showed that for a two-way contingency table the radii length of a $100(1 - \alpha)\%$ confidence circle for the ith row category in a two-dimensional correspondence plot is

$$r_{i(\alpha)}^I = \sqrt{\frac{\chi_\alpha^2}{np_{i\bullet}}} \tag{2.21}$$

where χ_α^2 is the $1 - \alpha$ percentile of a chi-squared distribution with two degrees of freedom. The basic idea is that if a confidence circle overlaps the origin (which, as we have discussed above, coincides with the position of all of the points under complete independence) then the category associated with that point does not contribute to the association structure, at the α level of significance. However, if a circle does not overlap the origin then the category that it visualises is deemed to make a statistically significant contribution to the association. This very simple idea has very important practical benefits for assessing what "close to the origin" means for a given row or column point and there have been more computationally intensive procedures proposed; see, for example, Greenacre (2017), Linting et al. (2007), Markus (1994) and Ringrose (1992, 1996). The problem with the confidence circles of Lebart et al. (1984) is that they do not take into consideration that the dimensions are weighted differently, nor do they capture all of the information contained in the third and higher dimensions.

To overcome these two problems, Beh (2010) proposed an adaptation of their confidence circles by considering elliptical regions, where the weight along each dimension helps to define the shape of the ellipse. Beh (2010) showed that the semi-major length, $x_{i(\alpha)}$, and the semi-minor axis length, $y_{i(\alpha)}$, for the ith row's $100(1 - \alpha)\%$ confidence ellipse is

$$x_{i(\alpha)} = \lambda_1 \sqrt{\frac{\chi_\alpha^2}{X^2} \left(\frac{1}{p_{i\bullet}} - \sum_{s=3}^{S} a_{is}^2 \right)} \tag{2.22}$$

and

$$y_{i(\alpha)} = \lambda_2 \sqrt{\frac{\chi_\alpha^2}{X^2} \left(\frac{1}{p_{i\bullet}} - \sum_{s=3}^{S} a_{is}^2 \right)}, \tag{2.23}$$

respectively, where X^2 is Pearson's chi-squared statistic of the two-way contingency table. These lengths reflect not just the association captured in the first two dimensions but they also take into account the information contained in the third and higher dimensions of a correspondence plot. Since $\lambda_1 > \lambda_2$ the semi-major axis length will always be parallel with the first dimension while the semi-minor axis length will always be parallel with the second dimension. Circular regions will arise if, and only if, $\lambda_1 = \lambda_2$. The presence of the summation term in Eq. (2.22) and Eq. (2.23) ensures that assessing the role played by a row or column point to the association between the variables is not just confined to the information contained in the two-dimensional display, but that it also reflects all of the information in the optimal correspondence plot.

If the ellipses are constructed to reflect only the association contained in the first two dimensions then the semi-major length and semi-minor lengths are

$$x_{i(\alpha)} = \sqrt{\frac{\lambda_1^2}{X^2/n} \frac{\chi_\alpha^2}{np_{i\bullet}}} \tag{2.24}$$

and

$$y_{i(\alpha)} = \sqrt{\frac{\lambda_2^2}{X^2/n} \frac{\chi_\alpha^2}{np_{i\bullet}}}, \tag{2.25}$$

respectively. Note that if the weight, λ_s, of each dimension is ignored (so that the term $\lambda_s^2/(X^2/n)$, for $s = 1,\ 2$ in Eq. (2.24) and Eq. (2.25) is ignored) then these two equations simplify to the radii length defined by Lebart et al. (1984); see Eq. (2.21).

More information on the implementation and interpretation of confidence circles constructed using Eq. (2.22) and Eq. (2.23) is given by Beh (2010) and Beh and Lombardo (2014).

2.9.2 Confidence Ellipses for the Traditional European Food Data

Recall that in Section 2.6 we described those *Free-word* and *Country* categories in Figure 2.6 that were close to, and far from, the origin. To assess whether those points far from the origin truly do not make a statistically significant contribution to the association between the variables, Figure 2.10 gives the 95% confidence ellipses for the row categories while

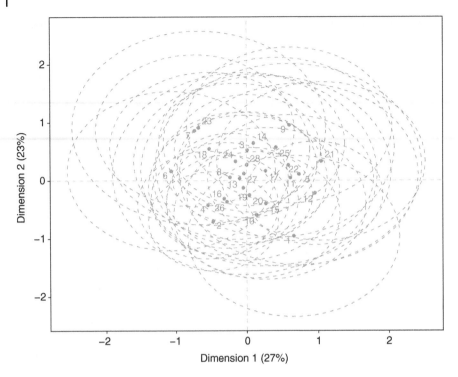

Figure 2.10 Correspondence plot of Table 1.1 with 95% confidence ellipses.

Figure 2.11 gives the same ellipses but for the column categories. The semi-major axis length and semi-minor axis length for the ellipses in both of these plots is calculated from Eq. (2.22) and Eq. (2.23), respectively. Figure 2.11 shows that all of the countries, except *Italy*, in Table 1.1 is a statistically significant contributor (at the $\alpha = 0.05$ level of significance) to the association between the variables. This is apparent since the confidence ellipse for all of the countries (except *Italy*) does not overlap the origin of the display.

Figure 2.10 also shows a very cluttered view of the ellipses for the rows of Table 1.1; we shall be addressing this issue in Section 2.10. Despite this, the 95% confidence ellipses show that all of the free-words listed in the contingency table, except one, do not make a statistically significant contribution to the association. The exception is *Habit* (Code 12) which appears to have its ellipse near, but not overlapping the origin. This suggests if the level of significance was any smaller than 0.05, say 0.01, the conclusion may differ. The lack of contribution to the association of the remaining free-words listed in Table 1.1 is apparent by observing that their 95% confidence ellipse overlaps the origin.

It also appears from Figure 2.10 and Figure 2.11 that the eccentricity of the ellipses is near circular; this is not surprising. To show why, we firstly note that Beh (2010) showed that the eccentricity of each confidence ellipse is identical (although the area may be quite different) and can be calculated by

$$E = \sqrt{1 - \left(\frac{\lambda_2}{\lambda_1}\right)^2}.$$

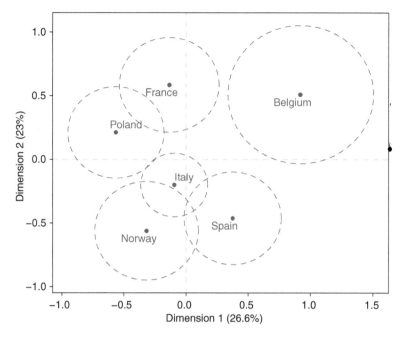

Figure 2.11 Correspondence plot of Table 1.1 with 95% confidence ellipses.

Here, E ranges between 0 and 1 (inclusive) where $E = 0$ means that the confidence region is a perfect circle. Therefore, since $\lambda_1^2 = 0.237$ and $\lambda_2^2 = 0.205$ (being somewhat similar in magnitude), the eccentricity of each of the confidence ellipses in Figure 2.10 and Figure 2.11 is

$$E = \sqrt{1 - \left(\frac{0.205}{0.237}\right)} = 0.367$$

and shows the near circular shape of the ellipses.

While the eccentricity of the ellipses remains constant for each of the row and column categories (for a given level of significance, α), their area is not constant since the semi-major and semi-minor axes are of different lengths for each category. For example, the area of the confidence ellipse for the ith row category is

$$A_\alpha(i) = \pi x_{i(\alpha)} y_{i(\alpha)}$$

$$= \pi \lambda_1 \lambda_2 \left[\frac{\chi_\alpha^2}{X^2}\left(\frac{1}{p_{i\bullet}} - \sum_{s=3}^{S} a_{im}^2\right)\right]$$

so that it depends on the row's marginal proportion, $p_{i\bullet}$ and the elements of its left singular vector. Table 2.10 gives the semi-major and minor length for each of the *Free-word* categories as well as the area of the ellipse for Table 1.1. It shows that the free-words that have the biggest elliptical area are *Quality* (Code 22) ($A_{0.05}(22) = 10.319$) and *Simple* (Code 27) ($A_{0.05}(27) = 10.244$). The large size of their ellipse is, in part, due to the very small number of classifications made in these rows of Table 1.1; note that 12 of the 743 sampled respondents associated *Quality* (Code 22) with traditional European food while for *Simple* (Code 27) its marginal frequency is 11. Therefore, there is insufficient evidence to conclude that

Table 2.10 Features of the 95% confidence ellipses for the *Free-word* categories listed in Table 1.1.

Code	Free-word	$x_{i(0.05)}$	$y_{i(0.05)}$	$A_{0.05}(i)$	$P_{i,2}$
1	Ancient	1.493	1.388	6.513	0.956
2	Christmas	0.919	0.854	2.466	0.141
3	Cooking	1.030	0.957	3.096	1.000
4	Country	0.998	0.928	2.910	1.000
5	Culture	1.681	1.563	8.257	1.000
6	Dinner	1.477	1.373	6.369	0.999
7	Dish	1.407	1.309	5.786	0.887
8	Family	0.817	0.760	1.952	1.000
9	Feast	1.445	1.344	6.102	0.942
10	Good	1.169	1.087	3.991	1.000
11	Grandmother	1.169	1.087	3.991	1.000
12	Habit	0.957	0.890	2.677	0.019
13	Healthy	1.227	1.141	4.400	1.000
14	Holidays	1.279	1.189	4.780	1.000
15	Home	1.405	1.307	5.769	1.000
16	Home-made	1.040	0.967	3.161	1.000
17	Kitchen	1.766	1.642	9.113	1.000
18	Meal	1.418	1.318	5.873	1.000
19	Natural	1.347	1.252	5.298	1.000
20	Old	0.885	0.823	2.288	1.000
21	Old-fashioned	1.469	1.366	6.304	0.998
22	Quality	1.880	1.748	10.319	1.000
23	Recipe	1.794	1.668	9.403	1.000
24	Regional	1.445	1.344	6.102	1.000
25	Restaurant	1.664	1.548	8.092	1.000
26	Rural	0.864	0.803	2.180	1.000
27	Simple	1.873	1.741	10.244	1.000
28	Tasty	0.983	0.914	2.822	1.000

these categories play a statistically significant role in defining the association between the two variables of Table 1.1. The row category with the smallest area of $A_{0.05}(8) = 1.952$ is *Family* (Code 8). Despite its small area, this row category still does not play a statistically significant role in defining the association. This is because, while its position in Figure 2.6 is very close to the origin, it has the highest marginal frequency of all of the free-words; the marginal frequency of *Family* (Code 8) is 63 and represents nearly 10% of the sample. Table 2.9 also shows that the correspondence plot of Figure 2.6 provides an excellent display of the category (thereby helping to verify that its close proximity to the origin is justified).

Table 2.11 Features of the 95% confidence ellipses for the
Country categories listed in Table 1.1.

Free-word	$x_{j(0.05)}$	$y_{j(0.05)}$	$A_{0.05}(j)$	$P_{j,2}$
Belgium	0.584	0.543	0.995	<0.001
France	0.400	0.372	0.468	<0.001
Italy	0.268	0.250	0.211	0.677
Norway	0.416	0.386	0.505	<0.001
Poland	0.386	0.359	0.436	<0.001
Spain	0.392	0.364	0.449	<0.001

Table 2.11 provides the same information as Table 2.10 but for each category of the *Country* variable. It shows that the area of the 95% confidence ellipses for each of the column categories of Table 1.1 is very small, at least in comparison to those summarised in Table 2.10 for the *Free-word* variable.

2.10 Approximate p-values

2.10.1 The Hypothesis Test and its p-value

There are sometimes situations when superimposing confidence circles, or confidence ellipses, onto a correspondence plot leads to a messy visualisation of the association. This is certainly the case with Figure 2.10. It may be apparent that the confidence ellipse for each free-word does not intersect the origin (if one looks very carefully) but it is not immediately clear exactly which of them overlaps the origin and which do not. To help remedy this problem p-values can be determined for each category. Such p-values provide a clearer impression when determining those row and column categories that contribute to the statistically significant symmetric association. Suppose we focus our attention for now on the row principal coordinates. Then, for the ith row category, its p-value is a reflection of the hypothesis test

$$H_0 : f_{is} = 0$$
$$H_A : f_{is} \neq 0$$

for each $i = 1, \ldots , I$ and $s = 1, \ldots , S$. For this test, Beh and Lombardo (2015) showed that the statistical significance of the ith row point in a two-dimensional correspondence plot can be assessed using the approximate p-value

$$P_{i,2} = P\left(\chi^2 > X^2 p_{i\bullet} \left(1 - \sum_{s=3}^{S} \left(\frac{f_{is}}{\lambda_s} \right)^2 \right)^{-1} \sum_{s=1}^{2} \left(\frac{f_{is}}{\lambda_s} \right)^2 \right). \tag{2.26}$$

The advantage of determining such a p-value is that it not only takes into account the information contained in a two-dimensional correspondence plot but it also reflects all of the information in the optimal correspondence plot; this can be seen with the inclusion of the $\sum_{s=3}^{S} (f_{is}/\lambda_s)^2$ term. If this term is ignored, the p-value reflects the rows information contained in only the first two dimensions. Generalisations of this p-value can be made if one wishes to construct a three-dimensional plot or consider the impact of a point in a more general $S^* \leq S$ dimensional correspondence plot.

2.10.2 P-values and the Traditional European Food Data

The last column of Table 2.10 gives the p-value of Eq. (2.26) for each of the *Free-word* categories of Table 1.1. These values reflect not just the position of each of the categories in this two-dimensional plot but they also take into account their position in the third and higher dimensions. It shows that all of the free-words, except *Habit* (Code 12), have a p-value that is larger than 0.05. In fact, of these free-words all, except *Christmas* (Code 2), have a p-value exceeding 0.8. This suggests that *Habit* (Code 12), with a p-value of 0.019, is the only *Free-word* category that contributes to the association between the two variables of Table 1.1. However, if the level of significance were increased to 0.10, say, then we would conclude that *Habit* (Code 12) does not play a statistically significant role in the association structure between the variables of Table 1.1. This confirms the comment we made of the 95% confidence ellipse of *Christmas* (Code 2) in Section 2.9.2.

To assess the contribution of each of the six countries of Table 1.1 to the statistically significant association between the two variables, the last column of Table 2.11 summarises their p-values, $P_{j,2}$. We can see here that, with the exception of *Italy*, all of the p-values are very small (being less than 0.001), and so all of the six countries (except *Italy*) contribute to the statistically significant symmetric association structure between *Country* and *Free-words* of Table 1.1. With a p-value of 0.677, *Italy* does not play a statistically significant role in the association and this is reflected by this country's confidence ellipse overlapping the origin in Figure 2.11; see also our comments on this in Section 2.9.2.

2.11 Final Comments

In this chapter we have examined some of the key features of simple correspondence analysis. In doing so, our attention has focused on providing a practical interpretation of the association between two symmetrically associated variables using a two-dimensional correspondence plot. The key word here is *symmetrically*. In the next chapter we shall discuss a similar strategy for visualising the association between categorical data but instead the focus is on studying? *asymmetrically* associated variables. Such a technique is referred to as *non-symmetrical correspondence analysis* and, as we shall see in Chapter 5, can be extended beyond this scenario.

3

Non-Symmetrical Correspondence Analysis

3.1 Introduction

In Chapter 2 we introduced the key features of the correspondence analysis of two categorical variables where both variables were treated as being symmetrically associated. However, such an association is not always the case. Sometimes there are many practical, or experimental, reasons where one may need to treat a variable as the predictor variable and the other as the response variable. While such a structure is very commonly taught in elementary statistics courses for numerical data (we only need to think about how simple linear regression is taught) very little attention (if any) is given to how to analyse the association structure between a categorical predictor variable and a categorical response variable. To help address this issue, the aim of this chapter is to provide an introduction to how to apply a correspondence analysis to a two-way contingency table where one variable is treated as a predictor variable and the other variable is treated as a response variable. Such a variant of correspondence analysis is referred to as *non-symmetrical correspondence analysis* and one may refer to Lauro and D'Ambra (1984) and D'Ambra and Lauro (1989) as its original sources. We shall confine our attention to the case where both variables consist of nominal categories. In Chapter 5 we extend our discussion to the non-symmetrical correspondence analysis of ordered categorical variables.

In Chapter 2, we applied simple correspondence analysis to the traditional European food data of Guerrero et al. (2010) that is summarised in Table 1.1. In doing so, we treated the *Free-word* and *Country* variables as both being predictor variables. Rather than treating the variables in this way, one may instead choose to study the association by treating the country of origin as a predictor of the choice of *Free-word* made to describe traditional European food. Therefore, in this chapter we explore the key issues of non-symmetrical correspondence analysis by focusing on this association structure. Our discussion of non-symmetrical correspondence analysis will be illustrated briefly by considering the generalised singular value decomposition (GSVD) of the table of weighted column profiles, where the weight takes into account the asymmetric association between the variables.

An Introduction to Correspondence Analysis, First Edition. Eric J. Beh and Rosaria Lombardo.
© 2021 John Wiley & Sons Ltd. Published 2021 by John Wiley & Sons Ltd.

3.2 Quantifying Asymmetric Association

3.2.1 The Goodman–Kruskal tau Index

When the two-way contingency table, $\mathbf{N}$, is formed from the cross-classification of a column predictor variable, Y, and a row response variable, X (note that here we are preserving the notation defined in Section 1.5.1), their *asymmetric association* should not be assessed using Pearson's chi-squared statistic. Instead one can use the Goodman–Kruskal tau index (Goodman and Kruskal 1954) to quantify this association. Due to the predictor/response nature of the association, the Goodman–Kruskal tau index is sometimes referred to as the *index of predictability*. We defined this index in Section 1.7.2 as

$$\tau = \frac{\sum_{i=1}^{I}\sum_{j=1}^{J}p_{\bullet j}\left(\frac{p_{ij}}{p_{\bullet j}} - p_{i\bullet}\right)^2}{1 - \sum_{i=1}^{I}p_{i\bullet}^2}$$

which is bounded to be between 0 and 1 (inclusive). The numerator of this index, denoted by

$$\tau_{\text{num}} = \sum_{i=1}^{I}\sum_{j=1}^{J}p_{\bullet j}\left(\frac{p_{ij}}{p_{\bullet j}} - p_{i\bullet}\right)^2,$$

is the weighted sum-of-squares of the elements of the centred column profiles. The denominator of τ quantifies the overall error in the prediction and does not depend on the predictor categories (D'Ambra and Lauro 1989; Lauro and D'Ambra 1984). We shall discuss τ and its numerator and denominator further in this section.

3.2.2 The τ Index and the Traditional European Food Data

The index, τ, is zero only when the profiles of all the predictor (column) categories are identical to the average column profile (under independence). In this case, there is no relative increase in the predictability of X given Y (Agresti 2002; Kroonenberg and Lombardo 1999). However this situation is quite rare in practice and small values of τ can be misleading. For example, for Table 1.1, $\tau_{\text{num}} = 0.033$ and $\tau = 0.035$ (to three decimal places) which suggests that there may be only a small level of "predictability" of the rows given the columns. However, such a small value is because we have a relatively large number of row categories ($I = 28$). While τ may appear to be small here, a formal test of the statistical significance of the predictability of X given Y can be made using the C-statistic defined by Eq. (1.8). For Table 1.1, this statistic is

$$C = (743 - 1)(28 - 1) \times 0.035$$
$$= 698.342$$

and is a chi-squared random variable with $(I - 1)(J - 1) = (28 - 1)(6 - 1) = 135$ degrees of freedom. By comparing this C-statistic with $\chi_{0.05}^2$ we find that the p-value is less than 0.001. Therefore, we can conclude that there is enough evidence to suggest that the country in which a respondent resides has a statistically significant influence on how they perceive traditional European food. For more details on the technical development of the C-statistic, refer to Light and Margolin (1971) and (Jung and Takane 2009b) for further information on testing procedures.

3.2.3 Weighted Centred Column Profile

Now that we have identified for Table 1.1 that the variable *Country* is a statistically significant predictor of the *Free-word* variable, we now explore in more detail the nature of this predictability. We do so by focusing on the numerator of the Goodman–Kruskal tau index which we express as

$$\tau_{\text{num}} = \sum_{i=1}^{I}\sum_{j=1}^{J}\tilde{c}_{i|j}^{2}$$

where

$$\tilde{c}_{i|j} = \sqrt{p_{\bullet j}}\left(\frac{p_{ij}}{p_{\bullet j}} - p_{i\bullet}\right). \tag{3.1}$$

Here, $\tilde{c}_{i|j}$ is the ith element of the weighted centred profile of the jth column when studying the asymmetric association between two categorical variables. This weighted centred profile is denoted by

$$\tilde{\mathbf{c}}_{j} = (\tilde{c}_{1|j},\ \ldots\ ,\ \tilde{c}_{I|j})$$

$$= \left(\sqrt{p_{\bullet j}}\left(\frac{p_{1j}}{p_{\bullet j}} - p_{1\bullet}\right),\ \ldots\ ,\ \sqrt{p_{\bullet j}}\left(\frac{p_{Ij}}{p_{\bullet j}} - p_{I\bullet}\right)\right).$$

Note that since the row variable is treated as a response variable it is considered to be a "random" variable. Therefore, we only focus our attention on comparing the centred column profiles using weights based on the predictor (column) variable. Recall that, in Chapter 2, when we treated the association to be symmetric the weights used were based on the row (predictor) and column (predictor) variables; see Eq. (2.2).

The denominator of the Goodman–Kruskal tau index, τ, which we express as

$$\tau_{\text{den}} = 1 - \sum_{i=1}^{I}p_{i\bullet}^{2}\ ,$$

plays no part in defining the association structure between the variables (other than for inferential purposes) since it only relies in the $p_{i\bullet}$ values and not p_{ij}. Instead, τ_{den} rescales τ_{num} so that it lies in the interval [0, 1]. Therefore, we define τ_{num} to be the measure of the total inertia when performing a non-symmetrical correspondence analysis on a two-way contingency table. More will be said on this measure in Section 3.3

3.2.4 Profiles of the Traditional European Food Data

Table 3.1 summarises the $\tilde{c}_{i|j}$ values of Table 1.1 where large positive values indicate that a *Country* is a good predictor of a *Free-word* when studying the perception of traditional

Table 3.1 The weighted centred column profiles, $\tilde{c}_j$, of Table 1.1.

Code	Free-word	Belgium	France	Italy	Norway	Poland	Spain
				Country			
1	Ancient	−0.008	−0.009	0.005	−0.009	−0.011	0.028
2	Christmas	−0.020	−0.021	−0.014	0.055	0.002	−0.005
3	Cooking	−0.008	0.037	−0.011	−0.017	−0.008	0.001
4	Country	−0.019	−0.009	−0.013	0.039	0.007	−0.011
5	Culture	0.008	0.001	−0.005	−0.008	−0.007	0.008
6	Dinner	−0.008	−0.009	−0.006	0.002	0.027	−0.012
7	Dish	−0.010	0.021	−0.007	−0.011	0.016	−0.014
8	Family	−0.019	0.017	0.006	0.002	0.002	−0.008
9	Feast	0.021	0.019	−0.006	−0.003	−0.013	−0.013
10	Good	−0.010	−0.009	0.023	0.003	−0.013	0.013
11	Grandmother	0.023	−0.004	−0.005	0.001	−0.015	0.003
12	Habit	0.032	−0.015	−0.015	−0.013	−0.030	0.036
13	Healthy	0.002	−0.012	0.007	−0.007	0.012	−0.002
14	Holidays	0.020	−0.012	−0.007	−0.011	0.019	−0.012
15	Home	−0.005	−0.007	−0.006	−0.010	0.001	0.019
16	Home-made	−0.015	−0.017	0.033	−0.009	0.013	0.003
17	Kitchen	0.009	−0.007	0.006	−0.003	0.002	−0.004
18	Meal	−0.006	0.008	−0.002	−0.011	0.016	−0.009
19	Natural	−0.002	−0.005	0.020	−0.004	0.000	−0.001
20	Old	−0.016	−0.001	−0.016	−0.004	−0.002	0.025
21	Old-fashioned	0.031	−0.008	−0.005	0.010	−0.010	−0.010
22	Quality	0.006	0.006	−0.004	−0.003	−0.008	0.002
23	Recipe	−0.006	0.016	−0.004	−0.007	0.008	−0.009
24	Regional	−0.004	0.014	0.016	−0.001	−0.008	−0.008
25	Restaurant	0.012	0.008	0.000	−0.001	−0.007	−0.008
26	Rural	−0.023	0.013	−0.011	0.042	−0.018	−0.003
27	Simple	0.002	−0.003	0.013	−0.002	0.001	−0.005
28	Tasty	0.011	−0.012	0.008	−0.021	0.020	−0.006

European food. For example, knowing that respondents are from *Spain* is helpful in predicting that they are (relatively speaking) more likely to associate the free-word *Ancient* (Code 1) with traditional European food than respondents from any of the other countries studied. This can be seen from Table 1.1; of the 190 Spanish respondents, 15/190 or about 8% of them associated traditional European food with *Ancient* (Code 1). There may be other words amongst those in *Spain* that have a higher frequency (including *Habit* (Code 12) and

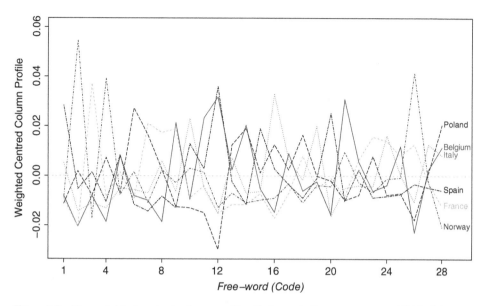

Figure 3.1 The weighted centred column profiles, $\tilde{c}_j$, for each *Country* category of Table 1.1.

Old (Code 20)) but if we knew that someone was from *Italy* (say) then only 2 out of 49, or about 4% of, respondents associated traditional European food with *Ancient* (Code 1).

From Table 3.1, we can also see that knowing that someone is from *Norway* is helpful in predicting that they are more likely to associate traditional European food with the free-word *Christmas* (Code 2). For those respondents in *Italy*, knowing where they come from helps to understand that they are highly likely to associate the free-words of *Home-made* (Code 16) and *Good* (Code 10) with traditional European food but that they do not perceive such food to be freely associated with *Old* (Code 20).

Figure 3.1 gives a simple visualisation of the weighted centred profile for each of the categories of the predictor (*Country*) variable. It shows that, given the *Country* in which a respondent resides, there are some free-words that are more frequently associated with traditional European food than others. The figure confirms those statements we just made when we discussed the magnitude of the values in Table 3.1.

Figure 3.1 also shows that a person from *Spain* or *Belgium* is highly likely to strongly associate traditional European food with the free-word *Habit* (Code 12), unlike those from *Poland* who do not perceive such food in this way. Furthermore, an individual from *Poland* is a good predictor of them associating traditional European food with *Dinner* (Code 6). The relatively flat profile of *Italy* suggests that knowing that someone is from this country is likely to only provide a weak prediction of almost all the free-words, except for *Good* (Code 10), *Home-made* (Code 16) and *Natural* (Code 19). However, the numerical summary (Table 3.1) and the visual comparison of the weighted centred profiles (see Figure 3.1) do not provide a complete view of the asymmetric association between *Free-word* and *Country*. Therefore, we shall perform a non-symmetrical correspondence analysis on Table 1.1 by treating *Country* as the predictor variable and *Free-word* as the response variable. Before we outline the application of this approach, we shall first provide an overview of its features.

3.3 Decomposing $\pi_{i|j}$ for Nominal Variables

3.3.1 The Generalised SVD of $\pi_{i|j}$

When numerically and visually summarising the asymmetric association between the column (predictor) and row (response) variables using non-symmetrical correspondence analysis, we apply the generalised singular value decomposition (GSVD) to the centred column profiles such that

$$\pi_{i|j} = \frac{p_{ij}}{p_{\bullet j}} - p_{i\bullet} = \sum_{s=1}^{S} a_{is} \lambda_s b_{js} . \tag{3.2}$$

Just as we described for simple correspondence analysis in Chapter 2, the maximum number of dimensions needed to visually summarise the predictability of the rows given the columns is $S = \min(I, J) - 1$. A feature of this GSVD is that, for each dimension, a row and column *score* can be obtained to visually summarise this predictability. For Eq. (3.2), the score for the ith row and jth column on the sth dimension (for $i = 1, 2, \ldots, I$ and $s = 1, 2, \ldots, S$) is a_{is} and b_{js}, respectively. These scores are the elements of the *left generalised vectors* and the *right generalised vectors*, respectively, and are normalised so that

$$\sum_{i=1}^{I} a_{is} = 0, \qquad \sum_{i=1}^{I} a_{is}^2 = 1 \tag{3.3}$$

$$\sum_{j=1}^{I} p_{\bullet j} b_{js} = 0, \qquad \sum_{j=1}^{J} p_{\bullet j} b_{js}^2 = 1 \tag{3.4}$$

for each of the S dimensions. These constraints imply that, from a statistical perspective, the expectation and variance of a_{is}, say, along the sth dimension is

$$E(a_{is}) = \sum_{i=1}^{I} a_{is}$$
$$= 0$$

and

$$\mathrm{Var}(a_{is}) = E(a_{is}^2) - [E(a_{is})]^2$$
$$= \sum_{i=1}^{I} a_{is}^2 - 0$$
$$= 1 .$$

Similarly, the expectation and variance of b_{js} is zero and one, respectively.

By defining $\pi_{i|j}$ as we have by Eq. (3.2), quantifying the absolute increase in predictability by the total inertia of the contingency table can be done by calculating the sum-of-squares of the singular values, λ_s, such that

$$\tau_{num} = \sum_{i=1}^{I} \sum_{j=1}^{J} p_{\bullet j} \pi_{i|j}^2$$
$$= \sum_{s=1}^{S} \lambda_s^2 . \tag{3.5}$$

Note that the sth singular value of the set of values $\pi_{i|j}$, λ_s, reflects the strength of the correlation between the right and left generalised singular vectors since

$$\lambda_s = \mathrm{Corr}(a_{is}, b_{js})$$

$$= \sum_{i=1}^{I}\sum_{j=1}^{J}p_{ij}\frac{(a_{is} - \mathrm{E}(a_{is}))}{\sqrt{\mathrm{Var}(a_{is})}}\frac{(b_{js} - \mathrm{E}(b_{js}))}{\sqrt{\mathrm{Var}(b_{js})}}$$

$$= \sum_{i=1}^{I}\sum_{j=1}^{J}p_{ij}a_{is}b_{js} \; .$$

One may define this singular value in terms of the $\pi_{i|j}$ values yielding the same quantity since

$$\lambda_s = \sum_{i=1}^{I}\sum_{j=1}^{J}\pi_{i|j}p_{\bullet j}a_{is}b_{js}$$

$$= \sum_{i=1}^{I}\sum_{j=1}^{J}p_{ij}a_{is}b_{js} \; . \tag{3.6}$$

3.3.2 GSVD and the Traditional Food Data

We now continue our study of the traditional European food data summarised in Table 1.1. Recall that for Table 1.1 we are treating *Country* as the predictor variable and *Free-word* as the response variable. Therefore, our first step to applying a non-symmetrical correspondence analysis on this data requires applying the GSVD on the matrix of $\pi_{i|j}$ values of Table 1.1 (summarised in Table 3.1). Doing so yields the squared singular values

$$\lambda_1^2 = 0.011, \quad \lambda_2^2 = 0.008, \quad \lambda_3^2 = 0.006, \quad \lambda_4^2 = 0.005, \quad \lambda_5^2 = 0.003$$

and are the explained inertia values from the decomposition. Therefore, using Eq. (3.5), the total inertia is

$$\tau_{\mathrm{num}} = \lambda_1^2 + \lambda_2^2 + \lambda_3^2 + \lambda_4^2 + \lambda_5^2$$

$$= 0.011 + 0.008 + 0.006 + 0.005 + 0.003$$

$$= 0.033$$

as expected, while the denominator of the Goodman–Kruskal tau index is $\tau_{\mathrm{den}} = 0.955$. Therefore, the Goodman–Kruskal tau index is

$$\tau = \frac{0.033}{0.955} = 0.035$$

which is the Goodman–Kruskal tau index we obtained in Section 1.7.3.

Multiplying the total inertia, τ_{num}, by the constant $(n-1)(I-1)/\tau_{\mathrm{den}}$ gives the C-statistic for Table 1.1 of

$$C = \frac{(743 - 1) \times (28 - 1) \times (0.011 + 0.008 + 0.006 + 0.005 + 0.003)}{0.955}$$

$$= 698.342$$

when τ_{num} and τ_{den} are calculated to three decimal places. This statistic can be compared with the critical value obtained from the chi-squared distribution with

Table 3.2 The a_{is} values from the GSVD of $\pi_{i|j}$ for Table 1.1.

Code	Free-word	Dim 1	Dim 2	Dim 3	Dim 4	Dim 5
1	Ancient	0.107	0.180	0.151	−0.353	0.069
2	Christmas	−0.548	0.241	0.188	0.152	0.092
3	Cooking	0.080	−0.104	−0.484	−0.236	0.095
4	Country	−0.440	0.074	0.065	0.135	0.112
5	Culture	0.123	0.098	−0.059	−0.023	0.061
6	Dinner	−0.119	−0.227	0.157	0.147	0.277
7	Dish	−0.034	−0.308	−0.229	−0.006	0.172
8	Family	−0.134	−0.176	−0.151	−0.167	−0.107
9	Feast	0.111	0.037	−0.329	0.227	−0.176
10	Good	−0.002	0.097	0.200	−0.277	−0.322
11	Grandmother	0.132	0.224	−0.048	0.189	−0.096
12	Habit	0.366	0.535	0.028	−0.012	0.161
13	Healthy	0.055	−0.115	0.198	0.061	0.035
14	Holidays	0.136	−0.129	0.115	0.362	0.202
15	Home	0.081	0.077	0.094	−0.185	0.270
16	Home-made	0.021	−0.215	0.408	−0.227	−0.228
17	Kitchen	0.070	−0.024	0.102	0.106	−0.086
18	Meal	0.007	−0.248	−0.053	0.005	0.146
19	Natural	0.038	−0.079	0.151	−0.087	−0.253
20	Old	−0.011	0.141	−0.010	−0.305	0.412
21	Old-fashioned	0.076	0.189	−0.015	0.419	−0.167
22	Quality	0.060	0.069	−0.119	−0.001	−0.016
23	Recipe	−0.019	−0.182	−0.179	−0.018	0.082
24	Regional	−0.003	−0.099	−0.116	−0.098	−0.354
25	Restaurant	0.060	0.017	−0.138	0.124	−0.153
26	Rural	−0.436	0.221	−0.225	−0.096	−0.135
27	Simple	0.032	−0.069	0.082	0.018	−0.185
28	Tasty	0.191	−0.225	0.218	0.148	0.092

$(28 - 1)(6 - 1) = 135$ degrees of freedom. With a p-value that is less than 0.001, there is ample evidence to suggest that the *Country* in which a person resides is a good predictor of the *Free-word* they are likely to associate with traditional European food.

By applying a non-symmetrical correspondence analysis to the data in Table 1.1, we can visualise the nature of the asymmetric association between its variables. We shall examine how to construct this visual summary, and its features, in Section 3.4.2. Before doing so, the numerical features from the GSVD of the $\pi_{i|j}$ values include its row and column singular vectors (the a_{is}'s and b_{js}'s respectively) and singular values (the λ_s's). Table 3.2 gives

Table 3.3 The b_{js} values from the GSVD of $\pi_{i|j}$ for Table 1.1.

Academic Level	Dim 1	Dim 2	Dim 3	Dim 4	Dim 5
Belgium	1.588	0.746	−0.152	1.925	−0.430
France	−0.013	−0.617	−2.041	−0.502	−0.316
Italy	0.501	−1.169	1.588	−1.058	−3.146
Norway	−2.009	1.009	0.123	0.638	−0.530
Poland	−0.224	−1.290	0.632	0.424	0.938
Spain	0.483	0.952	0.380	−1.120	0.610

the values of a_{is} from this decomposition which satisfy the constraints given by Eq. (3.3). Similarly, Table 3.3 summarises the b_{js} values from the GSVD of $\pi_{i|j}$ and these values satisfy the constraints of Eq. (3.4). Note that the optimal non-symmetrical correspondence plot consists of $S = \min(28, 6) - 1 = 5$ dimensions. Since there are $I = 28$ words in Table 1.1, there are 28 elements in each of the five left singular vectors while there are six elements in each of the five right singular vectors.

In Chapter 2 we described various strategies for visualising the symmetric association between the row and column variables of a two-way contingency table. These same approaches can also be adopted when there exists an asymmetric association between the variables, as we shall now discuss.

3.4 Constructing a Low-Dimensional Display

3.4.1 Standard Coordinates

Recall that for classical correspondence analysis, we apply the generalised singular value decomposition to the Pearson residuals – see Eq. (2.7) – to visualise the symmetric association between two categorical variables. We may also adapt this approach for asymmetrically associated variables by applying the GSVD to the centred column profiles – see Eq. (3.2) – using the elements of the left and right generalised vectors as coordinates for each of the row and column points of the contingency table. Such coordinates are referred to as *standard coordinates* for non-symmetrical correspondence analysis. Like those we discussed in Section 2.5.1, there are problems with using standard coordinates to visually summarise the asymmetric association between our two variables. Therefore, we shall not discuss this option further here. Instead our focus, just as it was in Chapter 2, is to focus on the interpretation of a correspondence plot constructed using *principal coordinates*.

3.4.2 Principal Coordinates

To visualise the asymmetric association between the variables of a two-way contingency table we define a *principal coordinate* for each category. The coordinate of the *i*th row category, and jth column category, along the sth dimension is defined as

$$f_{is} = a_{is}\lambda_s \tag{3.7}$$

$$g_{js} = b_{js}\lambda_s , \tag{3.8}$$

respectively. These principal coordinates are related to the $\pi_{i|j}$ values by

$$\pi_{i|j} = \frac{p_{ij}}{p_{\bullet j}} - p_{i\bullet} = \sum_{s=1}^{S} a_{is}\lambda_s b_{js} = \sum_{s=1}^{S} \frac{f_{is}g_{js}}{\lambda_s}$$

and shows that when f_{is} and g_{js} are zero along each dimension of the optimal correspondence plot (consisting of exactly S dimensions), for all $i = 1, \ldots , I$ and $j = 1, \ldots , J$ then there is no increase in predictability of the row variable given the column variable. Note that this is equivalent to complete independence since, for the (i, j)th cell of the contingency table, $\pi_{i|j} = 0$ only when $p_{ij} = p_{i\bullet}p_{\bullet j}$.

We can also show that the row and column principal coordinates defined by Eq. (3.7) and Eq. (3.8), respectively, are centred at the origin of the optimal correspondence plot by considering the following argument. Suppose we consider the row (response) principal coordinates. Then, since the information contained in the row categories is "not given", the expectation of f_{is} is simply

$$E(f_{is}) = \sum_{i=1}^{I}\sum_{s=1}^{S} f_{is}$$

$$= \sum_{i=1}^{I}\sum_{s=1}^{S} a_{is}\lambda_s$$

$$= \sum_{s=1}^{S} \lambda_s \left(\sum_{i=1}^{I} a_{is} \right)$$

$$= 0 .$$

Similarly, the variance of the row (response) principal coordinates is equivalent to the total inertia of the contingency table since

$$\text{Var}(f_{is}) = E(f_{is}^2) - [E(f_{is})]^2$$

$$= \sum_{i=1}^{I}\sum_{s=1}^{S} f_{is}^2 - 0$$

$$= \sum_{i=1}^{I}\sum_{s=1}^{S} a_{is}^2\lambda_s^2$$

$$= \sum_{s=1}^{S} \lambda_s^2 \left(\sum_{i=1}^{I} p_{i\bullet}a_{is}^2 \right)$$

$$= \sum_{s=1}^{S} \lambda_s^2$$

$$= \tau_{num}$$

using Eq. (3.5). Similar derivations can also be made to show that the column (predictor) principal coordinates are centred around the origin of the correspondence plot and their variation can also be measured using the total inertia, τ_{num}. For more insight into this feature, Beh and Simonetti (2011) provide an interpretation of these, and higher order, moments of the principal coordinates from the non-symmetrical correspondence analysis of a two-way contingency table.

These results show that, when performing non-symmetrical correspondence analysis, column (predictor) categories that are close to the origin suggest that they are not a good predictor of the row (response) categories. On the other hand, column (predictor) categories that lie far from the origin indicate that they are very good predictors. Like we described for simple correspondence analysis in Section 2.9, how "close" or "far" from the origin these points are in a non-symmetrical correspondence plot can be assessed using elliptical regions and p-values; more on these two aspects of the analysis will be described in Section 3.7.

Another important feature of the principal coordinates from a non-symmetrical correspondence analysis is their distance from the origin and from each other. For example, suppose we consider the ith row principal coordinate along the sth dimension of the optimal correspondence plot. Then the squared Euclidean distance of this point from the origin is

$$d_I^2(i,\,0) = \sum_{s=1}^{S} (f_{is} - 0)^2$$

$$= \sum_{s=1}^{S} f_{is}^2 \,.$$

Therefore, since

$$\tau_{num} = \sum_{i=1}^{I} \sum_{s=1}^{S} f_{is}^2$$

(see the derivation of $\mathrm{Var}(f_{is})$) the total inertia from this analysis can be expressed in terms of this distance since

$$\tau_{num} = \sum_{i=1}^{I} d_I^2(i,\,0) \,.$$

Therefore, row principal coordinates that are close to the origin of the optimal correspondence plot do not play an important role in defining the asymmetric association structure between the row and column variables. Conversely, points that lie far from the origin show that their categories do make such a contribution. We can demonstrate this same feature for the column principal coordinates since

$$\tau_{num} = \sum_{j=1}^{J} p_{\bullet j} d_J^2(j,\,0)$$

where

$$d_J^2(j,\,0) = \sum_{s=1}^{S} (g_{js} - 0)^2$$

$$= \sum_{s=1}^{S} g_{js}^2 \,.$$

We can also show that the squared Euclidean distance between two row (or column) principal coordinates, defined by Eq. (3.7) (or Eq. (3.8)) preserves the *property of distributional equivalence*. For example, suppose we consider the ith and i'th row profiles. Then, the distance between them in the optimal correspondence plot is expressed by

$$d_I^2(i,\ i') = \sum_{s=1}^{S} (f_{is} - f_{i's})^2$$

or in terms of the difference between their centred profiles so that

$$d_I^2(i,\ i') = \sum_{j=1}^{J} p_{\bullet j} \left(\frac{p_{ij}}{p_{i\bullet}} - \frac{p_{i'j}}{p_{i'\bullet}} \right)^2$$

$$= \sum_{j=1}^{J} p_{\bullet j} \left[\left(\frac{p_{ij}}{p_{i\bullet}} - p_{\bullet j} \right) - \left(\frac{p_{i'j}}{p_{i'\bullet}} - p_{\bullet j} \right) \right]^2 .$$

So, if the two row profiles are very similar to one another, then they will share a similar position to each other in the optimal correspondence plot. On the other hand, two profiles that are different will mean that they will be positioned at some distance away from each other in this plot. Note that we can also make similar comments about the comparison of two column profiles and their distance from each other in the optimal correspondence plot, but we shall leave this for the reader to ponder.

The key thing to note here is that we are speaking about the *optimal* correspondence plot. Therefore, just because two row points (say) are close to each other in a two-dimensional plot, this does not mean that their profiles will be very similar. We shall highlight this issue by comparing the configuration of points in Figure 3.3 and Figure 3.4 in the next section.

3.5 Practicalities of the Low-Dimensional Plot

3.5.1 The Two-Dimensional Correspondence Plot

Figure 3.2 gives the two-dimensional correspondence plot from applying a non-symmetrical correspondence analysis to Table 1.1; we refer to this plot as the *non-symmetrical correspondence plot*. Here, the row (response) and column (predictor) principal coordinates are defined by Eq. (3.7) and Eq. (3.8), respectively. The first thing that should be apparent from this figure is that the row principal coordinates are clustered very closely to the origin. This is because the elements of the left singular vectors are weighted differently to those described for simple correspondence analysis. This is apparent by comparing the relative magnitude of the values summarised in Table 3.2 and Table 3.3. So, to help provide a clearer view of the asymmetric association between the *Free-word* and *Country* variables we rescale the row and column principal coordinates given by Eq. (3.7) and Eq. (3.8), respectively, so that

$$\tilde{f}_{is} = \beta f_{is} \tag{3.9}$$

$$\tilde{g}_{js} = \frac{g_{js}}{\beta} \tag{3.10}$$

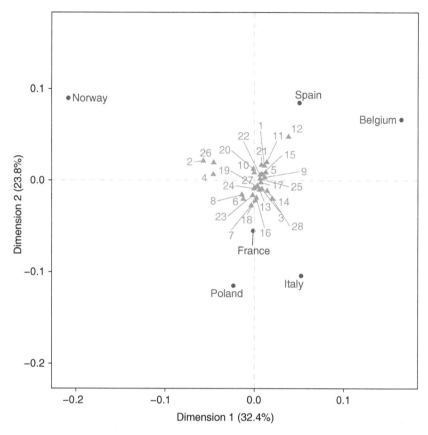

Figure 3.2 Two-dimensional non-symmetrical correspondence plot of Table 1.1.

for some value of β. Rescaling the principal coordinates in this way is akin to the *lambda scaling* described for biplot displays in Gower et al. (2011, Section 2.3.1). Due to the presence of β here we refer to this scaling as *beta scaling*. Using this form of scaling does not impact upon the general configuration of the row principal coordinates (only that they may appear further spread out from the origin for $\beta > 0$). Nor does it impact upon the general configuration of the column principal coordinates (other than their distance from the origin). Figure 3.3 provides a clearer indication of the association between each predictor (column) and response (row) category. It has been constructed by applying a beta scaling to the row and column principal coordinates where $\beta = 0.5$. For example, recall that a close observation and comparison of the weighted centred column profiles summarised in Table 3.1 and visualised in Figure 3.1 shows that all of the countries provide differing predictions of how their residents perceive traditional European food. Figure 3.3 shows that all countries have a different perspective on such food since their position in the correspondence plot varies.

Figure 3.3 also shows that knowing that someone resides from *Norway* predicts that they will associate traditional food with the free-words *Christmas* (Code 2), *Country* (Code 4) and *Rural* (Code 26) more so than any other word listed in Table 1.1. We can also see that knowing that somebody is from *Spain* or *Belgium* is a good predictor for them associating

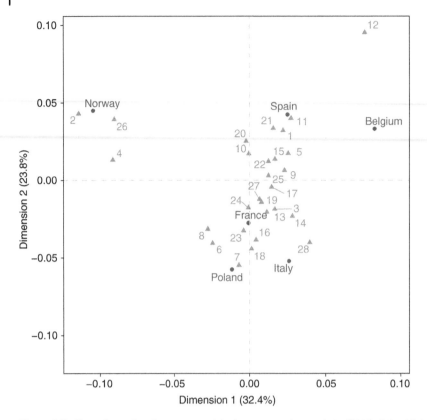

Figure 3.3 Two-dimensional non-symmetrical correspondence plot of Table 1.1 with beta scaling ($\beta = 0.5$).

traditional European food with the free-word *Habit* (Code 12). Indeed, *Spain* appears to be an excellent predictor of, for example, *Grandmother* (Code 11), *Ancient* (Code 1) and *Old-fashioned* (Code 21). Similarly, knowing that someone is from *Poland* is a good predictor of the free-words *Dish* (Code 7) and *Dinner* (Code 6) while a respondent from *France* is a good predictor of *Recipe* (Code 23) and *Regional* (Code 24).

We now turn our attention to the quality of Figure 3.3 for visualising the asymmetric association between the two variables of Table 1.1. Recall that the optimal correspondence plot consists of $S = 5$ dimensions. The first two of these dimensions visually describes only

$$100 \times \frac{0.011 + 0.008}{0.033} = 56.229\%$$

of the asymmetric association between *Country* and *Free-word*. Note that the first two explained inertia values, λ_1^2 and λ_2^2, are given here to three decimal places while this percentage is calculated using their values to five decimal places. Therefore, Figure 3.3 provides a better quality representation of the association than the symmetric correspondence plot of Figure 2.6 does (49.50%). However, one must be aware that the nature of the association depicted between the two plots is very different. This 56.229% comprises of

$$100 \times \frac{0.011}{0.033} = 32.401\%$$

of the total inertia, τ_{num} that is reflected by the first dimension. Similarly, the second dimension reflects

$$100 \times \frac{0.008}{0.033} = 23.828\%$$

of the asymmetric association. The third, fourth and fifth dimensions of the optimal correspondence plot account for 18.735%, 15.604% and 9.432%, respectively, of the asymmetric association between *Country* and *Free-word*. The first two percentage contributions to the total inertia are shown on the correspondence plot of Figure 3.3. However, with the plot only showing a relatively low proportion (less than 60%) of the total inertia, it is considered a poor visual summary of the asymmetric association.

We now discuss the impact of including a third dimension to the two-dimensional plot of Figure 3.3 and the information that is NOT being visually presented in this visual summary of the asymmetrical association.

3.5.2   The Three-Dimensional Correspondence Plot

Since a little less than 60% of the asymmetric association between the two variables of Table 1.1 is summarised in Figure 3.3, we shall now examine whether each of the row and column categories are well represented in the plot.

Recall that the total inertia obtained from performing a non-symmetrical correspondence analysis on the data in Table 1.1 is $\tau_{num} = 0.033$. For the resulting two-dimensional plot given by Figure 3.3 the first two dimensions account for

$$\lambda_1^2 + \lambda_2^2 = 0.011 + 0.008 = 0.019$$

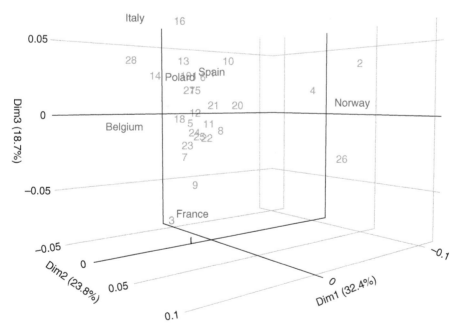

Figure 3.4   Three-dimensional non-symmetrical correspondence plot of Table 1.1 with beta scaling ($\beta = 0.5$).

of this inertia. Therefore, the third, fourth and fifth dimensions of the optimal correspondence plot account for

$$\lambda_3^2 + \lambda_4^2 + \lambda_5^2 = 0.006 + 0.005 + 0.003 = 0.014$$

of this inertia. We shall now examine how each free-word and country of Table 1.1 contributes to this 0.014. Large contributions will show that a category is very well described by the last three dimensions and so is poorly represented in Figure 3.3, while small contributions will indicate a very good representation of the category using the first two dimensions.

Following on from the arguments we made in Chapter 2, we assess where Figure 3.3 fails to adequately represent the asymmetric association by first reconstituting the $\pi_{i|j}$ values by

$$\hat{\pi}_{i|j} = \sum_{s=1}^{2} a_{is} \lambda_s b_{js} = \frac{f_{i1} g_{j1}}{\lambda_1} + \frac{f_{i2} g_{j2}}{\lambda_2}$$

where $\hat{\pi}_{i|j}$ is the approximation of $\pi_{i|j}$ given only the information in the first two dimensions. Here, f_{is} and g_{js} are defined by Eq. (3.7) and Eq. (3.8), respectively, although the beta scaled version of the coordinates, given by Eq. (3.9) and Eq. (3.10) respectively, can also be used without changing the following quantities. Therefore, determining

$$\tilde{\delta}_{ij} = \sqrt{P_{\bullet j}} (\pi_{i|j} - \hat{\pi}_{i|j})$$

$$= \sqrt{P_{\bullet j}} \left(\pi_{i|j} - \left(\frac{f_{i1} g_{j1}}{\lambda_1} + \frac{f_{i2} g_{j2}}{\lambda_2} \right) \right)$$

will provide some insight into where the two-dimensional non-symmetrical correspondence plot of Figure 3.3 fails to capture the asymmetric association between the *Country* and *Free-word* variables. By defining $\tilde{\delta}_{ij}$ in this manner, its sum-of-squares is

$$\tau_{\text{num (3:S)}} = \sum_{i=1}^{I} \sum_{j=1}^{J} \tilde{\delta}_{ij}^2$$

and is that part of the total inertia not captured in the first two dimensions.

Table 3.4 summarises the $\tilde{\delta}_{ij}$ values from the non-symmetrical correspondence analysis of Table 1.1. The smaller the value, the better the two-dimensional plot is at visualising the asymmetric association reflected in the $\pi_{i|j}$ value while large values reflect a poor visual summary; those 15 values in bold are the largest of the $\tilde{\delta}_{ij}$ values (ignoring the sign). Table 3.4 suggests that *France* and *Spain* are the two most poorly represented countries in Figure 3.3. This is the case because of the relatively high number of large (in bold) values of $\tilde{\delta}_{ij}$. However, to make such an observation neglects the remaining values of $\tilde{\delta}_{ij}$ in each country.

We can quantify the contribution that each free-word and each country makes to the total inertia of the third and higher dimensions. Such contributions are calculated by

$$\tilde{\delta}_{I(i)}^2 = \sum_{j=1}^{J} \tilde{\delta}_{ij}^2$$

for the ith free-word so that the total inertia in these dimensions is calculated by

$$\tau_{\text{num (3:S)}} = \sum_{i=1}^{I} \tilde{\delta}_{I(i)}^2 \ .$$

Table 3.4 The $\tilde{\delta}_{ij}$ values of Table 1.1.

Code	Free-word	Country					
		Belgium	France	Italy	Norway	Poland	Spain
1	Ancient	**−0.018**	−0.005	0.008	−0.007	0.000	**0.018**
2	Christmas	0.006	−0.015	−0.001	0.003	0.009	−0.002
3	Cooking	−0.010	**0.035**	−0.015	−0.007	−0.013	0.004
4	Country	0.005	−0.007	−0.005	0.001	0.007	−0.003
5	Culture	−0.002	0.004	−0.004	−0.001	0.000	0.001
6	Dinner	0.004	−0.015	−0.010	0.000	0.013	0.001
7	Dish	−0.001	0.014	−0.014	−0.003	−0.001	−0.001
8	Family	−0.007	0.013	0.003	−0.002	−0.009	0.003
9	Feast	0.014	**0.020**	−0.007	0.005	−0.009	**−0.017**
10	Good	−0.012	−0.007	**0.026**	0.000	−0.007	0.009
11	Grandmother	0.010	0.001	−0.001	0.004	−0.001	−0.010
12	Habit	−0.002	−0.003	−0.006	−0.002	0.005	0.004
13	Healthy	0.002	−0.014	0.004	0.001	0.007	0.001
14	Holidays	0.016	−0.015	−0.012	0.005	0.013	−0.009
15	Home	−0.012	−0.005	−0.005	−0.006	0.006	0.014
16	Home-made	−0.011	**−0.022**	**0.027**	0.000	0.001	0.012
17	Kitchen	0.006	−0.008	0.005	0.003	0.002	−0.004
18	Meal	−0.001	0.002	−0.008	−0.002	0.003	0.001
19	Natural	−0.003	−0.007	**0.017**	0.002	−0.004	0.001
20	Old	**−0.019**	0.002	−0.013	−0.010	0.006	**0.019**
21	Old-fashioned	**0.022**	−0.004	−0.001	0.009	0.002	**−0.020**
22	Quality	0.001	0.008	−0.003	0.000	−0.003	−0.002
23	Recipe	−0.001	0.012	−0.009	−0.002	−0.002	−0.001
24	Regional	−0.001	0.012	0.014	0.002	−0.013	−0.004
25	Restaurant	0.008	0.009	0.000	0.003	−0.005	−0.010
26	Rural	−0.003	**0.018**	0.001	−0.001	−0.010	−0.002
27	Simple	0.002	−0.004	0.010	0.002	−0.002	−0.003
28	Tasty	0.005	**−0.017**	0.000	0.002	0.010	−0.001

Similarly, the contribution of the jth country to the total inertia of the third and higher dimensions is quantified by

$$\tilde{\delta}^2_{J(j)} = \sum_{i=1}^{I} \tilde{\delta}^2_{ij}$$

so that

$$\tau_{\text{num (3:S)}} = \sum_{j=1}^{J} \tilde{\delta}^2_{J(j)} \,.$$

Table 3.5 Contribution ($\tilde{\delta}_{I(i)}^2$) and percentage contribution, (%) of each *Free-word* category to the third and higher dimensions of the optimal non-symmetrical correspondence plot.

	Free-word (Code)			
Quantity	**Ancient (1)**	**Christmas (2)**	**Cooking (3)**	**Country (4)**
$\delta_{I(i)}^2$	0.0004	0.0004	0.0018	0.0002
%	5.479	2.740	12.329	1.370
	Culture (5)	Dinner (6)	Dish (7)	Family (8)
$\delta_{I(i)}^2$	<0.0001	0.0005	0.0004	0.0003
%	–	3.425	2.740	2.055
	Feast (9)	Good (10)	Grandm. (11)	Habit (12)
$\delta_{I(i)}^2$	0.0010	0.0010	0.0002	0.0001
%	6.849	6.849	1.370	0.685
	Healthy (13)	Holidays (14)	Home (15)	Home-made (16)
$\delta_{I(i)}^2$	0.0003	0.0009	0.0005	0.0015
%	2.055	6.164	3.425	10.274
	Kitchen (17)	Meal (18)	Natural (19)	Old (20)
$\delta_{I(i)}^2$	0.0001	0.0001	0.0004	0.0010
%	0.685	0.685	2.740	6.849
	Old-fash. (21)	Quality (22)	Recipe (23)	Regional (24)
$\delta_{I(i)}^2$	0.0010	0.0001	0.0002	0.0005
%	6.849	0.685	1.370	3.425
	Restaurant (25)	Rural (26)	Simple (27)	Tasty (28)
$\delta_{I(i)}^2$	0.0003	0.0004	0.0002	0.0004
%	2.055	2.740	1.370	2.740

The $\tilde{\delta}_{J(j)}^2$ values are summarised, to three decimal places, in the first row of Table 3.6. The percentage contribution of each of these values relative to $\tau_{num\,(3:S)} = 0.014$ is summarised in the second row. It shows that *France* accounts for more than a third of the total inertia described by the third and higher dimensions of the optimal correspondence plot; this conclusion is consistent with our observation of the most extreme values of $\tilde{\delta}_{ij}$ summarised in Table 3.4. However, we also see that *Italy* also contributes to a relatively large proportion of this total inertia. The countries of *Norway* and *Poland* contribute very little to these higher dimensions and so is best represented in the two-dimensional correspondence plot of Figure 3.3.

By following a similar strategy to that described in the previous paragraph, we can also identify how well each word (or row) of Table 1.1 is represented in Figure 3.3. Table 3.5 summarises the $\tilde{\delta}_{I(i)}^2$ values and their percentage contribution to $\tau_{num\,(3:S)} = 0.014$ for each free-word of Table. 1.1. It shows that *Cooking* (Code 3) and *Home-made* (Code 16) are the two largest contributors to the total inertia in the third and higher dimensions. Together they account for a third (or 33%) of $\tau_{num\,(3:S)}$ and so there is a lot of information they contain

Table 3.6 Contribution ($\tilde{\delta}^2_{J(j)}$) and percentage contribution, (%) of each *Country* category to the third and higher dimensions of the optimal non-symmetrical correspondence plot.

	Country						
Quantity	Belgium	France	Italy	Norway	Poland	Spain	Total
$\tilde{\delta}^2_{J(j)}$	0.002	0.005	0.003	0.001	0.001	0.002	0.014
%	14.286	35.713	21.429	7.143	7.143	14.286	100.000

about the asymmetric association between *Free-word* and *Country* not accounted for in Figure 3.3. This suggests that adding an additional dimension to the two-dimensional correspondence plot would provide an improved visual summary of the association, especially for these free-words. Doing so would also improve the depiction of the contribution *France* makes to this association. In fact, by adding a third dimension to Figure 3.3, thereby yielding Figure 3.4, improves the visualisation of the association by a further 18.735%. Viewing the third dimension of Figure 3.4 shows that free-words such as *Cooking* (Code 3) and *Home-made* (Code 16) have a relatively large coordinate along the third dimension of the plot. So too does *France* and *Italy* all of which help to confirm our comments above. Note that Figure 3.4 is constructed by applying the beta scaling ($\beta = 0.5$) to the row and column principal coordinates; see Eq. (3.9) and Eq. (3.10).

3.6 The Biplot Display

3.6.1 Definition

An alternative visual representation of the asymmetric association between the *Free-word* and *Country* variables of Table 1.1 is to construct a biplot instead of the classic non-symmetrical correspondence plot; recall that these correspondence plots were given by Figure 3.3 (using the first two dimensions) and Figure 3.4 (using the first three dimensions) using beta scaling with $\beta = 0.5$. The construction of the biplot for non-symmetrical correspondence analysis follows the same approach as its construction for simple correspondence analysis that we described in Section 2.7. That is, the *i*th row and *j*th column biplot coordinate along the *s*th dimension is defined by

$$\tilde{f}_{is} = a_{is}\lambda_s^\epsilon \tag{3.11}$$

$$\tilde{g}_{js} = b_{js}\lambda_s^{1-\epsilon}, \tag{3.12}$$

respectively, where, typically, $0 \le \epsilon \le 1$. The difference between these formulae and the biplot coordinates used in simple correspondence analysis – see Eq. (2.19) and Eq. (2.20) – is that the elements of the left *s*th singular vector, $\{a_{is} : i = 1, 2, \ldots, I\}$ in Eq. (3.11), are subject to the constraint given by Eq. (3.3) while, for the symmetric analysis, the singular vector is constrained by Eq. (2.8). The constraints imposed upon the elements of the right *s*th singular vector $\{b_{js} : j = 1, 2, \ldots, J\}$ in the column biplot coordinates defined by Eq. (3.12) are those given by Eq. (3.4).

Just as we described in Section 2.7, a column isometric biplot is constructed by simultaneously visualising the row standard coordinates and the column principal coordinates obtained from a non-symmetrical correspondence analysis of a two-way contingency table. This arises when $\epsilon = 0$ yielding a biplot where each row category is depicted as a projection from the origin to their position defined by their standard coordinate while each column category is depicted as a point at their principal coordinate. Similarly, a row isometric biplot is constructed by simultaneously plotting the row principal coordinates and the column standard coordinates (projected from the origin), when $\epsilon = 1$.

3.6.2 The Column Isometric Biplot for the Traditional Food Data

Since we are treating the *Country* of Table 1.1 as the predictor variable and *Free-word* as the response variable, we construct a column isometric biplot to visualise their asymmetric association; see Figure 3.5. Therefore, row (response) categories are depicted as projections from the origin to their position in the plot (with beta scaling applied, $\beta = 1.5$) while the column (predictor) categories are depicted simply as points.

By observing the configuration in Figure 3.5, we can see that *France* is positioned close to the origin. Therefore, just like we saw when we were interpreting the two-dimensional non-symmetrical correspondence plot of Figure 3.3, it appears that knowing that a respondent is from *France* is a poor predictor of how they perceive traditional European

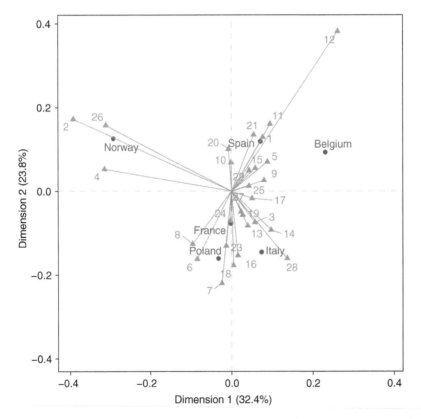

Figure 3.5 Column isometric biplot from the non-symmetrical correspondence analysis of Table 1.1 with beta scaling ($\beta = 1.5$).

food. Figure 3.5 also shows that knowing that someone is from *Norway* is a good predictor that they will associate traditional European food with the free-words *Christmas* (Code 2), *Country* (Code 4) and *Rural* (Code 26). Furthermore, given that an individual is from *Spain*, it is highly likely that they associate traditional European food with the free-words *Ancient* (Code 1), *Grandmother* (Code 11) and *Old-fashioned* (Code 21). Similarly, given a respondent from *Belgium*, we can infer that they identify traditional European food with, for example, *Feast* (Code 9) and *Restaurant* (Code 25) while those from *Italy* associate such food with *Tasty* (Code 28), *Healthy* (Code 13), *Natural* (Code 19) and *Simple* (Code 27). Suppose now that we know an individual comes from *Poland*. Figure 3.5 shows that they generally associate the free-words *Dinner* (Code 6), *Dish* (Code 7) and *Family* (Code 8) with traditional European food. Similarly, knowing that an individual is from *France*, implies that they associate *Regional* (Code 24), *Home-made* (Code 16) and *Meal* (Code 18), for example, with traditional European food.

3.6.3   The Three-Dimensional Biplot

Recall that when applying a non-symmetrical correspondence analysis to the data in Table 1.1, 56.229% of the total inertia ($\tau_{num} = 0.033$) is reflected in the two-dimensional column isometric biplot given by Figure 3.5. To help improve the quality of the visual summary, we can add a third dimension to the biplot. Doing so adds a further 18.735% to the quality of the display and is given by Figure 3.6; note that there are no row

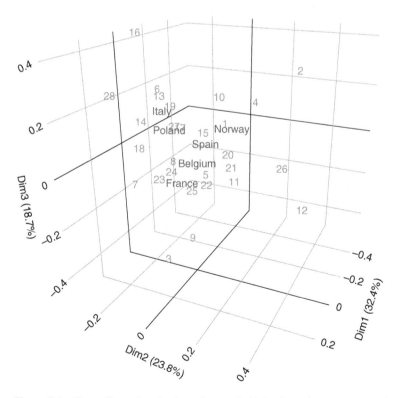

Figure 3.6   Three-dimensional column isometric biplot from the non-symmetrical correspondence analysis of Table 1.1 with beta scaling ($\beta = 1.5$).

projections here. Therefore, this three-dimensional biplot visually summarises 74.964% of the asymmetric association between the variables *Free-words* and *Country*. There are some differences in the findings provided by Figure 3.5 and Figure 3.6. For example, Figure 3.6 suggests that a respondent from *Norway* is not as likely to associate traditional European food with *Rural* (Code 26) as the two-dimensional biplot of Figure 3.5 suggests. Similarly, Figure 3.5 suggests a strong asymmetric association between *Italy* and *Cooking* (Code 3) and yet the three-dimensional biplot suggests a very poor association since the distance of the free-word from the projection of *Italy* is great. We could perform a detailed investigation of what a two-dimensional biplot (or correspondence plot) is NOT representing, like we did for symmetrically associated variables in Section 2.7.3 (see also Section 2.6.2), but we shall leave this for the interested reader to pursue.

3.7 Detecting Statistically Significant Points

3.7.1 Confidence Circles and Ellipses

We now turn our attention to assessing how "close" to the origin a point is in a two-dimensional non-symmetrical correspondence plot. Beh (2010) showed that the radii length of a $100(1 - \alpha)\%$ confidence circle for the jth column (predictor) category, say, is

$$
r^J_{j(\alpha)} = \sqrt{\frac{\chi^2_\alpha \left(1 - \sum_{i=1}^I p_{i\bullet}^2\right)}{p_{\bullet j}(n-1)(I-1)}}
\tag{3.13}
$$

where χ^2_α is the $1 - \alpha$ percentile of the chi-squared distribution with two degrees of freedom. Lombardo et al. (2007) also derived this length for the non-symmetrical correspondence analysis of a two-way contingency table formed from the cross-classification of an ordered row variable and an ordered column variable. Like the confidence circles described for simple correspondence analysis (see Section 2.9), confidence circles with a radii length of Eq. (3.13) that overlap the origin highlight those categories that do not make a statistically significant contribution (at the α level of significance) to the asymmetric association between the variables. On the other hand, confidence circles that do not overlap the origin identify those categories that are statistically significant contributors to this association.

Since $\lambda_1 > \lambda_2 > \ldots > \lambda_S$, the axes are not equally weighted and confidence circles with a radii length given by Eq. (3.13) ignores this feature of the correspondence plot. The construction of the circles also ignore the association reflected in the third and higher dimensions. To overcome these two issues, Beh and Lombardo (2015) derived confidence ellipses along the same lines as those presented in Section 2.9. They showed that the semi-major length, $\tilde{x}_{i(\alpha)}$, and the semi-minor axis length, $\tilde{y}_{i(\alpha)}$, for the ith row's $100(1 - \alpha)\%$ confidence ellipse is

$$
\tilde{x}_{i(\alpha)} = \lambda_1 \sqrt{\frac{\chi^2_\alpha \left(1 - \sum_{i=1}^I p_{i\bullet}^2\right)}{\tau_{num}(n-1)(I-1)} \left(1 - \sum_{s=3}^S a_{is}^2\right)}
\tag{3.14}
$$

and

$$\tilde{y}_{i(\alpha)} = \lambda_2 \sqrt{\frac{\chi_\alpha^2 \left(1 - \sum_{i=1}^I p_{i\bullet}^2\right)}{\tau_{num}(n-1)(I-1)} \left(1 - \sum_{s=3}^S a_{is}^2\right)}, \tag{3.15}$$

respectively. One may also refer to Beh (2010), Lombardo and Ringrose (2012) and Beh and Lombardo (2015) for additional comments and insights into the construction and interpretation of confidence regions for correspondence analysis.

Two important features of the confidence ellipse are its eccentricity and area. Firstly, just like the confidence ellipses described in Section 2.9.2 for simple correspondence analysis, the confidence ellipse for non-symmetrical correspondence analysis have a constant eccentricity for all categories at all levels of significance. This eccentricity is calculated as

$$E = \sqrt{1 - \left(\frac{\lambda_2}{\lambda_1}\right)^2}.$$

The second important feature of these confidence circles is their area. For example, the area of the $100(1 - \alpha)\%$ confidence ellipse for the ith row (response) category from the non-symmetrical correspondence analysis of a two-way contingency table is

$$\tilde{A}_\alpha(i) = \pi \lambda_1 \lambda_2 \left[\frac{\chi_\alpha^2 \left(1 - \sum_{i=1}^I p_{i\bullet}^2\right)}{\tau_{num}(n-1)(I-1)} \left(1 - \sum_{s=3}^S a_{is}^2\right)\right]. \tag{3.16}$$

3.7.2 Confidence Ellipses for the Traditional Food Data

Consider again Table 1.1. Figure 3.7 gives the non-symmetrical correspondence plot of Figure 3.2 with the 95% confidence ellipses superimposed. This figure represents only the row (response) categories while Figure 3.8 provides the same plot but for the column (predictor) categories. Given the large number of row categories ($I = 28$ words), there is a lot of overlap of the row ellipses and so it is difficult to discern those categories that are well predicted and those that are not. However, a careful inspection of Figure 3.7 shows that most of the confidence ellipses overlap the origin. The exceptions are the ellipses of the free-words *Christmas* (Code 2), *Rural* (Code 26) and *Habit* (Code 12). Therefore, we can conclude from this figure that only these free-words are well predicted given the *Country* variable at the 5% level of significance. The remaining free-words are not deemed to be statistically significant contributors to the asymmetric association between the two variables of Table 1.1 since their ellipses overlap the plots origin. In fact, by calculating the p-value of the row (response) categories by

$$\tilde{P}_{i,2} = P\left\{\chi^2 > \frac{\tau_{num}(n-1)(I-1)}{1 - \sum_{i=1}^I p_{i\bullet}^2} \left(\frac{1}{p_{i\bullet}} - \sum_{s=3}^S a_{is}^2\right)^{-1} \sum_{s=1}^S \left(\frac{f_{is}}{\lambda_s}\right)^2\right\}$$

we see that most of the free-words have a very large p-value suggesting that they are not well predicted given the *Country* variable. Those free-words that are, statistically speaking,

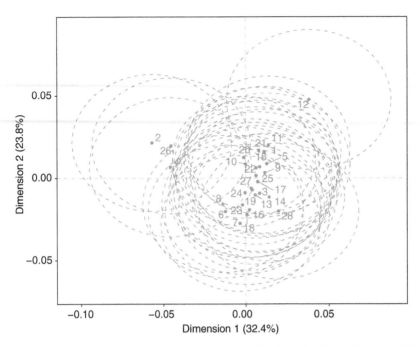

Figure 3.7 Non-symmetrical correspondence plot of Table 1.1 with 95% (row) confidence ellipses.

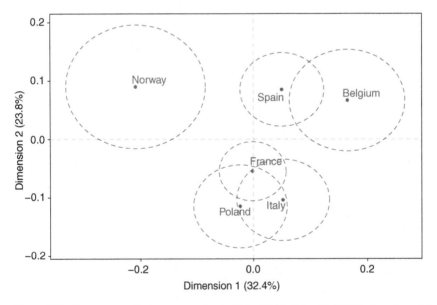

Figure 3.8 Non-symmetrical correspondence plot of Table 1.1 with 95% (column) confidence ellipses.

Table 3.7 Features of the 95% confidence ellipses for the categories of the *Free-word* variable of Table 1.1.

Code	Free-word	$\tilde{x}_{i(0.05)}$	$\tilde{y}_{i(0.05)}$	$\tilde{A}_{0.05}(i)$	$\tilde{P}_{i,2}$
1	Ancient	0.046	0.040	0.006	1.000
2	Christmas	0.048	0.042	0.006	<0.001
3	Cooking	0.042	0.036	0.005	1.000
4	Country	0.049	0.042	0.007	0.276
5	Culture	0.050	0.043	0.007	1.000
6	Dinner	0.047	0.040	0.006	1.000
7	Dish	0.048	0.041	0.006	1.000
8	Family	0.049	0.042	0.006	1.000
9	Feast	0.045	0.039	0.005	1.000
10	Good	0.044	0.038	0.005	1.000
11	Grandmother	0.049	0.042	0.006	1.000
12	Habit	0.050	0.042	0.007	<0.001
13	Healthy	0.049	0.042	0.006	1.000
14	Holidays	0.045	0.039	0.006	1.000
15	Home	0.047	0.040	0.006	1.000
16	Home-made	0.043	0.037	0.005	1.000
17	Kitchen	0.049	0.042	0.007	1.000
18	Meal	0.050	0.043	0.007	1.000
19	Natural	0.048	0.041	0.006	1.000
20	Old	0.043	0.037	0.005	1.000
21	Old-fashioned	0.045	0.038	0.005	1.000
22	Quality	0.050	0.043	0.007	1.000
23	Recipe	0.049	0.042	0.007	1.000
24	Regional	0.046	0.040	0.006	1.000
25	Restaurant	0.049	0.042	0.006	1.000
26	Rural	0.048	0.041	0.006	0.005
27	Simple	0.049	0.042	0.007	1.000
28	Tasty	0.048	0.041	0.006	1.000

well predicted are *Christmas* (Code 2), *Habit* (Code 12) and *Rural* (Code 26); the first two row (response) categories have a p-value that is less than 0.001 while the p-value of *Rural* (Code 26) is 0.005. These p-values are summarised in the last column of Table 3.7 and provide a numerical justification for the conclusions reached by observing each category's 95% confidence ellipse.

An inspection of Figure 3.8 shows that the confidence ellipse of each of the categories of the *Country* variable does not overlap the origin. Therefore, we can conclude that all six

Table 3.8 Features of the 95% confidence ellipses for the categories of the *Country* variable of Table 1.1.

Country	$\tilde{x}_{j(0.05)}$	$\tilde{y}_{j(0.05)}$	$\tilde{A}_{0.05}(j)$	$\tilde{p}_{j,2}$
Belgium	0.101	0.087	0.028	<0.001
France	0.059	0.051	0.009	0.001
Italy	0.081	0.070	0.018	<0.001
Norway	0.123	0.106	0.041	<0.001
Poland	0.123	0.106	0.041	<0.001
Spain	0.073	0.063	0.015	<0.001

countries are statistically significant predictors of the *Free-word* categories of Table 1.1. The country whose ellipse that comes closest to the origin of the plot is that of *France*. Therefore, one may suspect that the p-value for this column (predictor) category is closest to, but still well less than, 0.05. The p-values for these column categories are summarised in the last column of Table 3.8 and shows that all of the countries, except *France*, have a p-value that is less than 0.001. However, the p-value associated with *France* is still very small (0.001) providing evidence that all six countries contribute to the asymmetric association between the two variables of Table 1.1.

We provide the following comments on the eccentricity and area of the 95% confidence ellipses in Figure 3.7 and Figure 3.8. Concerning their eccentricity, since the first two explained inertia values are $\lambda_1^2 = 0.011$ and $\lambda_2^2 = 0.008$, the eccentricity of the confidence ellipses from performing a non-symmetrical correspondence analysis on the data in Table 1.1 is

$$E = \sqrt{1 - \frac{0.008}{0.011}} = 0.522 .$$

We note here that the ellipses for non-symmetrical correspondence analysis are much less circular than the ellipses derived from a simple correspondence analysis of Table 1.1. This is because, for simple correspondence analysis, λ_2^2 is about 16% bigger than λ_1^2 while for non-symmetrical correspondence analysis λ_2^2 is about 38% bigger than λ_1^2.

Concerning the area of the row (response) category ellipses, they are calculated using Eq. (3.16) – a similar equation can be used for the columns. The area of the 95% confidence ellipse for the row (response) categories are summarised in the fifth column of Table 3.7 while the area of the ellipse for the column (predictor) categories are summarised in Table 3.8.

3.8 Final Comments

In this chapter, we have extended our discussion of the features and application of simple correspondence analysis by focusing on studying the asymmetric association structure between the categorical variables of a two-way contingency table. In doing so, we have presented an overview of non-symmetrical correspondence analysis; sometimes the terminology appears in the literature as *non symmetric correspondence analysis, non-symmetric correspondence analysis* or *non symmetrical correspondence analysis* (note the inclusion and exclusion of "-" here). While our discussion has been concerned with defining the column variable as the predictor variable and the row variable as the response variable, the same features and comments apply if one were to study a contingency table where the row variable is defined as the predictor variable and the column variable is defined as the response variable. For the purposes of performing a non-symmetrical correspondence analysis in this case, the analysis can be carried out by simply transposing the contingency table being studied.

An important aspect of our discussion in this chapter, and of the discussion made in the "symmetric" case, in Chapter 2, is that we have treated the two variables as consisting of nominal categories. In many practical situations, this is not always the case since a variable may well include ordered categories. Simple examples of this include any variable measured on a Likert scale, or gradations of eye colour, hair colour, income level, age, etc. Therefore, in Chapter 4 and Chapter 5 we shall extend the discussion given in the last two chapters by describing the simple and non-symmetrical correspondence analysis techniques for ordered categorical variables. Like we have done in this chapter and Chapter 2 our discussion of ordered variables will be restricted to the study of only two categorical variables. In Chapter 4 we will highlight how simple correspondence analysis can be amended for the case where both variables consist of ordered categories. Chapter 5 will focus the non-symmetrical correspondence analysis of a two-way contingency table consisting of ordered column (predictor) categories and nominal row (response) categories.

Part II

Ordinal Analysis of Two Categorical Variables

4

Simple Ordinal Correspondence Analysis

4.1 Introduction

Chapter 2 introduced some of the key features of simple correspondence analysis when the variables consist of nominal categories. In many practical settings though, categorical variables consist of ordered categories. It is therefore appropriate to discuss how the simple correspondence analysis technique described in Chapter 2 can be amended to capture this additional structure.

There are various ways that one can take into account the "orderedness" of such variables when performing a simple correspondence analysis. For example, one can refer to Ritov and Gilula (1983), Schriever (1983), Parsa and Smith (1993) and Yang and Huh (1999) for a description of various approaches. Others to have considered the issue of ordered categorical variables in the context of correspondence analysis include Nishisato and Arri (1975), Nishisato (1980, 2007), Nair (1986), Gifi (1990), Heiser and Meulman (1994), Beh (1997), Meulman et al. (2004) and Agresti (2010). For row and column variables with an increasing ordered structure, many of these approaches, broadly speaking, maximise the correlation, λ_s, between a_{is} and b_{js} subject to the additional constraint that $a_{1s} < a_{2s} < \ldots < a_{Is}$ and $b_{1s} < b_{2s} < \ldots < b_{Js}$ for some s. Usually $s = 1$ is set so that the row and column inertias are ordered along the first dimension of a correspondence plot. Such a strategy assumes that the most dominant correlation between the two ordered categorical variables is one that describes the linear-by-linear association. However, it may be that the most dominant source of association is non-linear in nature.

The correspondence analysis approach that we shall describe in this chapter is based on the technique described in Beh (1997) for a doubly ordered contingency table; where both variables of a two-way table are ordinal. This approach captures linear and non-linear sources of association that may exist between the variables. We shall briefly outline how this variant of correspondence analysis can be performed for row and column variables that both consist of ordered categories. For those interested in how this technique can be amended for a singly ordered contingency table, refer to Beh (2001a) and Lombardo and Beh (2016). Their approach involves a hybridisation of the decomposition method introduced in this chapter for the analysis of ordered categorical variables and generalised singular value decomposition (GSVD).

When a two-way contingency table is formed by cross-classifying one, or two, ordered categorical variables the strategy that we take for decomposing the (i, j)th Pearson residual,

An Introduction to Correspondence Analysis, First Edition. Eric J. Beh and Rosaria Lombardo.
© 2021 John Wiley & Sons Ltd. Published 2021 by John Wiley & Sons Ltd.

$\gamma_{ij} - 1$, is different to what is traditionally adopted for nominal variables. From a practical perspective, the structure of ordinal variables is often neglected when a correspondence analysis is performed; the exceptions, of course, include those mentioned above. Indeed, even Chapter 2 and Chapter 3 provided no guidance as to how the correspondence analysis of ordinal categorical variables can be performed. This is primarily because the presence of an ordered variable adds an extra source of complexity when finding a solution to determining row scores and column scores while maximising the association between them.

So how can we determine such scores while maximising the correlation between the variables?

In Section 4.4 we shall describe an alternative strategy to the GSVD described and applied in Chapter 2 and Chapter 3. The strategy that we adopt is to decompose the Pearson residuals of a two-way contingency table using what we call *bivariate moment decomposition* (shortened here to BMD). Such a variant of correspondence analysis may be referred to as *doubly ordered correspondence analysis*, or simply as DOCA; see Beh and Lombardo (2014, Chapter 6) for additional details of this variant.

The case where a two-way contingency table consists of a nominal categorical variable and an ordered categorical variable often arises in many practical settings; the variant of correspondence analysis that can be performed in this case is referred to as *singly ordered correspondence analysis*, or SOCA. We shall not discuss this variant of ordered correspondence analysis when the variables are symmetrically associated. However, we do provide an overview of how such a simple correspondence analysis approach can be performed when a contingency table consists of a nominal categorical variable and an ordinal categorical variable when the variables are asymmetrically associated. Such a technique is referred to as *singly ordered non-symmetrical correspondence analysis*, or SONSCA, and is described in Chapter 5.

For more information on how to perform an ordered correspondence analysis on a doubly ordered contingency table, and some of the interesting numerical features that form a basis of this variant, we direct the reader to Beh (1997, 2001a, 2008), Rayner and Beh (2009), Rayner and Best (2000) and Beh and Lombardo (2014, Section 6.7).

4.2 A Simple Correspondence Analysis of the Temperature Data

Consider the 1973 temperature data summarised in Table 1.2. A chi-squared test of independence reveals that the chi-squared statistic is 104.31 with 12 degrees of freedom. With a p-value less than 0.001, there is enough evidence to conclude that there is an association between the month of the year (May through to September) and the temperature recorded at La Guadia airport in New York in 1973. It would be surprising if we did not detect this association since, meteorologically, it is to be expected that the temperature at the airport would change as the seasons change.

While both variables, *Month* and *Temperature*, are ordinal, often such a feature is ignored and the contingency table is analysed using the classical approach to correspondence analysis which assumes that the variables are nominal. So before we perform a doubly ordered correspondence analysis on Table 1.2, we shall first turn our attention to performing the classical (nominal) approach described in Chapter 2.

If the two variables are treated as being nominal and we apply a classical two-way correspondence analysis then $S = \min(4, 5) - 1 = 3$, and

$$\lambda_1^2 = 0.520, \qquad \lambda_2^2 = 0.126, \qquad \lambda_3^2 = 0.041 .$$

Therefore, the total inertia of the data is

$$\frac{X^2}{n} = 0.520 + 0.126 + 0.041$$
$$= 0.687 .$$

A visual summary of the association can be made using the two-dimensional correspondence plot given by Figure 4.1. It suggests that *May* is the coldest month and is associated
with temperatures in the range of (56, 72] degrees Fahrenheit. It also suggests that the
month of *July* is not the hottest, that honour belongs to *August*; *July* appears to have a
strong interaction with the temperature range (79, 85] degrees while, for *August*, this range
is (85, 97].

Figure 4.1 shows that the lowest temperature range, (56, 72], is associated with the month
of *May* suggesting that, in the period of study, it is consistently the coldest month. It also
shows that the hottest month, for (85, 97], is *August* while the two hottest months, with

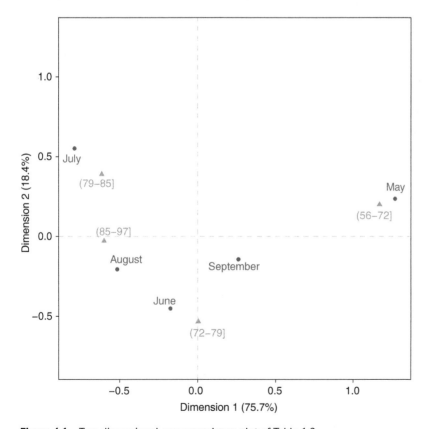

Figure 4.1 Two-dimensional correspondence plot of Table 1.2.

a temperature range of (79, 97], are *July* and *August*. The most temperate month, with a temperature range of (72, 79] is *June*.

Figure 4.1 also gives a very good visual representation of the association, reflecting

$$100 \times \frac{\lambda_1^2 + \lambda_2^2}{X^2/n} = 100 \times \frac{0.520 + 0.126}{0.687}$$

$$= 94.1\%$$

of the total inertia, but one must be a little cautious about whether it is a fair graphical summary of the association since there are clearly large, and not-so large, cell frequencies that dominate the configuration. This is because a feature of the classical approach to simple correspondence analysis is that a row and column point is very much influenced by the magnitude of the cell frequency they share. For example, *May* and (56, 72] has the largest cell frequency in Table 1.2 of 24 and it is this value, in part, that influences their close proximity to each other in Figure 4.1. Similarly, *August* and (85, 97] have a relatively large cell frequency of 14 thereby drawing their points close to one another in the correspondence plot.

It is also unclear how the profiles of the row and column categories differ. For example, it is not specified how the profiles of *May* and *July*, which are located on opposite sides of Figure 4.1, are different; it only shows that they are different. One may also question how the profiles of the maximum and minimum temperature ranges of Table 1.2 differ. Certainly one can imply that the impact of these profiles plays a different role in defining the association structure between the two variables, but it does not provide any guidance as to what their role is.

We can address these issues and accommodate the structure of the ordered variables using the approach to simple correspondence analysis described in this chapter. However, before we describe this variant, we first examine how ordered categories can be incorporated into the analysis by using ordered row scores and ordered column scores.

4.3 On the Mean and Variation of Profiles with Ordered Categories

4.3.1 Profiles of the Temperature Data

In Chapter 2 we examined the features of simple correspondence analysis by first comparing the row profiles and the column profiles of Table 1.1. Therefore, to help explore the features of the simple correspondence analysis of the doubly ordered table of Table 1.2, the *i*th centred row profile is defined by

$$\tilde{\mathbf{r}}_i = \left(\frac{p_{i1}}{p_{i\bullet}} - p_{\bullet 1}, \ldots, \frac{p_{iJ}}{p_{i\bullet}} - p_{\bullet J} \right).$$

The centred row profiles of Table 1.2 are summarised in the rows of Table 4.1. Similarly, the *j*th centred column profile is defined by

$$\tilde{\mathbf{c}}_j = \left(\frac{p_{1j}}{p_{\bullet j}} - p_{1\bullet}, \ldots, \frac{p_{Ij}}{p_{\bullet j}} - p_{I\bullet} \right).$$

Table 4.1 The centred row profiles of Table 1.2.

Temperature	Month					
(F)	May	June	July	August	September	Total
(56, 72]	0.434	−0.118	−0.204	−0.178	0.066	0.000
(72, 79]	−0.075	0.168	−0.155	0.016	0.047	0.000
(79, 85]	−0.172	−0.018	0.283	−0.024	−0.069	0.000
(85, 97]	−0.197	−0.050	0.090	0.208	−0.050	0.000

Table 4.2 The centred column profiles of Table 1.2.

Temperature	Month				
(F)	May	June	July	August	September
(56, 72]	0.550	−0.150	−0.250	−0.218	0.083
(72, 79]	−0.103	0.230	−0.205	0.021	0.064
(79, 85]	−0.223	−0.023	0.356	−0.031	−0.090
(85, 97]	−0.224	−0.057	0.099	0.228	−0.057
Total	0.000	0.000	0.000	0.000	0.000

The centred column profiles of Table 1.2 are summarised in the columns Table 4.2.

A visual summary of these profiles is also presented. Figure 4.2 shows the centred row profiles. It shows that there is indeed variation in the location (or mean) and dispersion (or variation) of the profiles across the five month period. For example, the smallest temperature range, (56, 72] is very much dominated in the month of *May* and drops considerably between *June* and *August* before increasing in *September*. Such a pattern is due to the changes in temperature during the summer months (in the Northern Hemisphere) where we can see that the temperature range (79, 85] peaks during *July*.

Figure 4.3 displays the centred column profiles. It shows, for example that the low temperature range is very much dominated by the month of *May* while the hottest temperature range is dominated by *July* and *August*.

4.3.2 Defining Scores

Suppose we treat the column variable as consisting of nominal categories. In doing so there is no ordered structure and the weights given to each column will be identical. So, define the uniform weight for the jth column by w. Then the "weighted" centred profile of the first row category is

$$w \times (0.434) + w \times (-0.118) + w \times (-0.204) + w \times (-0.178) + w \times (0.066) = 0 .$$

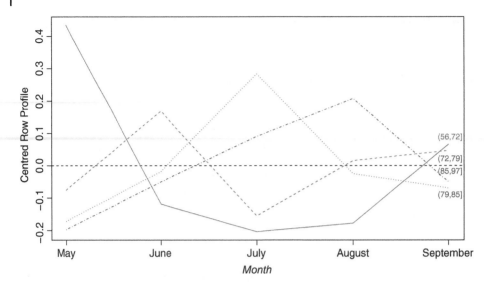

Figure 4.2 Centred row profiles, $\tilde{r}_i$, of Table 1.2.

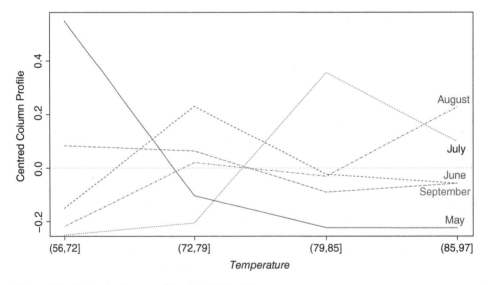

Figure 4.3 Centred column profiles, $\tilde{c}_j$, of Table 1.2.

This result shows that, if the categories are treated as being nominal then the mean of this profile is at zero.

However, since there is an ordered structure to the *Month* variable, assigning equal weights to each of its categories does not prove useful. Instead we can assign "weights" to the column categories so that these "weights" reflect the ordered structure of the variables. Rather than referring to them as "weights" we shall adopt the terminology *scores* so that, in general, the jth column category is assigned a score of $s_I(j)$. Therefore, we can assess the weighted mean of the first row profile using these scores so that

$$s_J(1)(0.434) + s_J(2)(-0.118) + s_J(3)(-0.204) + s_J(4)(-0.178) + s_J(5)(0.066) \ .$$

Similarly, we denote the score assigned to the ith row category by $s_I(i)$. The vector of row scores, $\mathbf{s}_I = (s_I(1), \ldots, s_I(I))^T$ and column scores $\mathbf{s}_J = (s_J(1), \ldots, s_J(J))^T$ may be referred to as the set of *basis scores*, as we shall see in Section 4.4.1.

A special case of these scores is when $s_I(i) = i$ and $s_J(j) = j$ for $i = 1, \ldots, I$ and $j = 1, \ldots, J$, respectively; such scores are referred to as *natural scores*. Therefore, by using these scores to reflect the ordered structure of the column variable, the mean of the first row category, (56, 72], shifts from zero to

$$1 \times (0.434) + 2 \times (-0.118) + 3 \times (-0.204) + 4 \times (-0.178) + 5 \times (0.066) = -0.796 \ .$$

The negative sign of this mean shows that this temperature range is dominated by the earlier months (such as *May* and/or *June* than at any other period). The problem though is that while the mean of the profile will differ for different row categories, this measure does not provide any information about the variation of the profile. We shall discuss more on this issue soon.

While natural scores imply that the adjacent categories are equi-distant in their (ordered) difference from one another, they remain a popular choice for many categorical data analysts. This is because they are not only simple to use but there are various methodological advantages when calculating and interpreting quantities that use them. We discuss a key advantage of their use from a simple correspondence analysis perspective in Section 4.4.1. Of course, one may adopt alternative scoring strategies, as long as the chosen set of scores adequately reflect the ordered structure of the variable and provide a meaningful interpretation of the numerical and graphical features of the analysis. One advantage in assigning *a priori* scores – as we have done by deciding on the choice of $\mathbf{s}_I$ and $\mathbf{s}_J$ – to ordered categories when performing correspondence analysis is in that there is a clear and meaningful interpretation of the mean and variation of a profile. Before we describe how ordered categorical variables are incorporated into a simple correspondence analysis we first discuss the role of these scores for assessing differences in the mean and variation of the set of row (or column) profiles.

4.3.3 On the Mean of the Profiles

To reflect any differences in the mean of the row profiles, keeping in mind that the columns are ordered, we define the mean of the ith ordered centred row profile by

$$\mu_I(i) = \sum_{j=1}^{J} s_J(j) \left(\frac{p_{ij}}{p_{i\bullet}} - p_{\bullet j} \right) = \sum_{j=1}^{J} s_J(j) \frac{p_{ij}}{p_{i\bullet}} - \mu_I \tag{4.1}$$

where μ_I is the overall mean row profile

$$\mu_I = \sum_{j=1}^{J} s_J(j) p_{\bullet j} \ . \tag{4.2}$$

If the order of the column categories is increasing (for example, quantities ranging from small to large or low to high) then the scores are selected so that $s_J(1) < s_J(2) < \ldots < s_J(J)$.

Similarly, for a set of increasing row scores, so that $s_I(1) < s_I(2) < \ldots < s_I(I)$, the mean of the jth ordered centred column profile is

$$\mu_J(j) = \sum_{i=1}^{I} s_I(i) \left(\frac{p_{ij}}{p_{\bullet j}} - p_{i\bullet} \right) = \sum_{i=1}^{I} s_I(i) \frac{p_{ij}}{p_{\bullet j}} - \mu_J \tag{4.3}$$

where

$$\mu_J = \sum_{i=1}^{J} s_I(i) p_{i\bullet} \tag{4.4}$$

is the overall mean column profile. While we will not be going into any great detail here, Eq. (4.2) and Eq. (4.4) are, respectively, very closely related to

$$a_{is} = \sum_{j=1}^{J} b_{js} \left(\frac{p_{ij}}{p_{i\bullet}} - p_{\bullet j} \right)$$

$$b_{js} = \sum_{i=1}^{I} a_{is} \left(\frac{p_{ij}}{p_{\bullet j}} - p_{i\bullet} \right).$$

These two formulae form the starting point for the method of reciprocal averaging, a method used to find the solution to a_{is} and b_{js} and gives identical results to those obtained from the GSVD of the Pearson residuals; see Eq. (2.7). For more details on the method of reciprocal averaging in the context of correspondence analysis, see for example, Hill (1974), Gifi (1990) and Beh and Lombardo (2014, Chapter 3).

4.3.4 On the Variation of the Profiles

One feature that the traditional approach to simple correspondence analysis (of nominal variables) does not consider is the difference between the profiles in terms of their variation. Some profiles exhibit a large amount of variation while others do not. So, to assess the differences in the variation of the row profiles, where the column variable consists of ordered categories, a simple measure of variation for the ith row profile is

$$\sigma_I^2(i) = \sum_{j=1}^{J} s_J(j) \left(\frac{p_{ij}}{p_{i\bullet}} - \mu_I(i) \right)^2 \tag{4.5}$$

while

$$\sigma_J^2(j) = \sum_{i=1}^{I} s_I(i) \left(\frac{p_{ij}}{p_{\bullet j}} - \mu_J(j) \right)^2 \tag{4.6}$$

is the measure of variation of the jth column profile.

4.3.5 Mean and Variation of Profiles for the Temperature Data

To demonstrate the calculation of $\mu_I(i)$ and $\mu_J(j)$ consider the two-way contingency table of Table 1.2. This table is formed from the cross-classification of two ordered categorical variables; *Temperature* which consists of four temperature ranges and *Month* which includes the period *May* through to *September*, for 1973. Since Table 1.2 consists of two ordered variables, we refer to it as a *doubly ordered two-way contingency table*, or more simply as a *doubly ordered table*. To assess the difference in the mean of the row profiles, we shall be using natural scores to reflect the ordinal structure of the column variable. That is we define the basis vector (or scores) for the column categories so that $\mathbf{s}_J = (1, 2, 3, 4, 5)$. To quantify the differences in the mean of the row profiles, we first find that

$$\mu_I = \sum_{j=1}^{J} s_J(j) p_{\bullet j}$$
$$= 1 \times \frac{30}{152} + 2 \times \frac{30}{152} + 3 \times \frac{31}{152} + 4 \times \frac{31}{152} + 5 \times \frac{30}{152}$$
$$= 3.007$$

using Eq. (4.2). Therefore, using Eq. (4.1) for $i = 1$, the mean of the first temperature category, (56, 72], is

$$\mu_I(1) = \sum_{j=1}^{J} s_J(j) \left(\frac{p_{1j}}{p_{1\bullet}} - p_{\bullet j} \right)$$
$$= 1(0.434) + 2(-0.118) + 3(-0.204) + 4(-0.178) + 5(0.066)$$
$$= -0.796$$

just as we described in Section 4.3.2. Similarly, the mean of the profile for each of the remaining temperature categories of Table 1.2 is

$$\mu_I(2) = 0.091, \qquad \mu_I(3) = 0.199, \qquad \mu_I(4) = 0.552 .$$

Therefore, the first row profile – for the temperature range (56, 72] – has a mean that is much less than the overall mean profile while the mean of the profile for the highest temperature range, (85, 97], is higher than the overall mean. In fact, the mean of the profile for the temperature range of (56, 72] is more extreme (when compared with the overall mean) than that of the temperature range (85, 97]. Of the four row categories used to describe the temperature ranges at La Guardia airport, the profile of the second temperature range, (72, 79] is very close to the mean row profile since $\mu_I(2)$ is close to zero.

We can make similar remarks for the five months studied. However, rather than using natural scores, we can instead use the mid-point of each interval to represent the difference between each of the adjacent temperature ranges. That is, we define the rows scores so that

$$\mathbf{s}_I = (s_I(1), s_I(2), s_I(3), s_I(4))$$
$$= \left(\frac{56 + 72}{2}, \frac{72 + 79}{2}, \frac{79 + 85}{2}, \frac{85 + 97}{2} \right)$$
$$= (64.0, 75.5, 82.0, 91.0) .$$

Therefore, using Eq. (4.4), the mean of all the column profile distributions is $\mu_J = 77.760$ so that the mean of each of the five monthly profiles is

$$\mu_J(1) = -11.243, \qquad \mu_J(2) = 0.690, \qquad \mu_J(3) = 6.724,$$

$$\mu_J(4) = 5.837, \qquad \mu_J(5) = -2.427 .$$

These values have been calculated using Eq. (4.3). We see here that, even though the order of the row categories has been reflected by using natural scores, the mean of the five column profiles does not preserve this order; note that for the months of *May* ($j = 1$) and *September* ($j = 5$) their mean profile is less than the overall mean.

We now examine the variation that exists between the row profiles and the column profiles. Using Eq. (4.5) for $i = 1$, we find that the variation of the first row profile can be quantified by

$$\sigma_I^2(1) = \sum_{j=1}^{J} s_J(j)\left(\frac{p_{1j}}{p_{1\bullet}} - \mu_I(1)\right)^2$$

$$= 1 \times \left(\frac{24}{38} - (-0.796)\right)^2 + 2 \times \left(\frac{3}{38} - (-0.796)\right)^2$$

$$+ 3 \times \left(\frac{0}{38} - (-0.796)\right)^2 + 4 \times \left(\frac{1}{38} - (-0.796)\right)^2$$

$$+ 5 \times \left(\frac{10}{38} - (-0.796)\right)^2$$

$$= 13.785$$

while the variation of the profiles for the remaining row categories is

$$\sigma_I^2(2) = 0.340, \qquad \sigma_I^2(3) = 0.307, \qquad \sigma_I^2(4) = 1.733 .$$

Similarly, the variation of the five column profiles are

$$\sigma_J^2(1) = 41041.864, \qquad \sigma_J^2(2) = 67.064, \qquad \sigma_J^2(3) = 13033.263,$$

$$\sigma_J^2(4) = 9699.987, \qquad \sigma_J^2(5) = 2225.926$$

using Eq. (4.6). These quantities show that the row profile with the greatest variation is for the temperature range (56, 72] with $\sigma_I^2(1) = 13.785$. This should not be surprising even if the ordering of the column categories is not reflected since this particular temperature range has cell frequencies ranging from the smallest count of 0 up to the largest count of 24; this variation can also be observed in Figure 4.2 where its profile is the least flat of the four temperature ranges. On the other hand, the row profile of the temperature range of (79, 85] has the least variation. If we focus now on the column profiles, the *Month* that exhibits the most variation is *May* which has cell frequencies ranging from a minimum of 0 to the maximum count of 24; we can also see from Figure 4.3 that the variation of the temperatures is certainly the most extreme during this period. Numerically, it also has the largest variation with $\sigma_J^2(1) = 41041.864$. The column profile with the least variation is for the month of *June*; it is during this month that the temperatures remain the most consistent with $\sigma_J^2(2) = 67.064$. One may see that its profile in Figure 4.3 is the most flat of the *Month* profiles and so reflects this relatively small variation.

Now that we have established how to quantify differences in the mean and variation of a set of row (or column) profiles for a doubly ordered contingency table, we now turn our attention to how these concepts can be incorporated into simple correspondence analysis.

4.4 Decomposing the Pearson Residual for Ordinal Variables

4.4.1 The Bivariate Moment Decomposition of $\gamma_{ij} - 1$

Recall that for the simple correspondence analysis of a two-way contingency table consisting of nominal variables, we perform a GSVD on the Pearson residuals; see Eq. (2.7). A feature of GSVD is that it does not easily accommodate the structure of ordered categorical variables. However, for a doubly ordered contingency table, one way to overcome this problem is to instead perform a *bivariate moment decomposition* (BMD) of the Pearson residuals. Mathematically, BMD looks very similar to SVD and is of the form

$$\gamma_{ij} - 1 = \sum_{u=1}^{I-1}\sum_{v=1}^{J-1}\alpha_{iu}\lambda_{uv}\beta_{jv} \,. \tag{4.7}$$

However, the interpretation of the resulting terms is very different to those obtained when performing a GSVD. In Eq. (4.7), α_{iu} is a uth order orthogonal polynomial for the ith row and β_{jv} is a vth order orthogonal polynomial for the jth column. These values can be found using the simple recurrence formulae of Emerson (1968) and have been extensively discussed in the correspondence analysis literature over the past 20 years; see Beh (1997) for the first use of these formulae in correspondence analysis. The interested reader may also consider Beh (1998, 1999, 2001a, 2004a,b, 2008), Lombardo et al. (2007), Lombardo and Beh (2010), Lombardo and Meulman (2010), Lombardo et al. (2011), Simonetti et al. (2011), Beh and Lombardo (2014), Lombardo and Beh (2016), and Lombardo et al. (2016a) for additional descriptions of their use in this context. Haberman (1974) also uses orthogonal polynomials for his derivation of ordinal log-linear models but uses those described by Kendall and Stuart (1967, pp. 356–361).

Suppose we consider the first-order row orthogonal polynomials α_{i1}, for $i = 1, 2, \ldots, I$. These values are akin (but not exactly equal) to the mean values $\mu_I(i)$, for $i = 1, 2, \ldots, I$, defined by Eq. (4.2). Similarly, the first-order column orthogonal polynomials β_{j1}, for $j = 1, 2, \ldots, J$ are akin (but not exactly equal) to the mean values $\mu_J(j)$, for $j = 1, 2, \ldots, J$, defined by Eq. (4.4). Where α_{i1} and $\mu_I(i)$ (and β_{j1} and $\mu_J(j)$) differ is in how they are constrained. The mean values calculated using Eq. (4.2) and Eq. (4.4) are unconstrained while the elements of the row (α_{iu}) and column (β_{jv}) orthogonal polynomials are constrained in a manner that is analogous to the constraints imposed on a_{im} and b_{jm} described in Section 2.4.1. That is

$$\sum_{i=1}^{I}p_{i\bullet}\alpha_{iu} = 0, \qquad \sum_{i=1}^{I}p_{i\bullet}\alpha_{iu}^2 = 1$$

for $u = 1, \ldots, I - 1$, while

$$\sum_{j=1}^{I}p_{\bullet j}\beta_{jv} = 0, \qquad \sum_{j=1}^{J}p_{\bullet j}\beta_{jv}^2 = 1$$

for $v = 1, \ldots, J - 1$. Simply speaking, determining these polynomials first requires specifying an initial set of *basis scores* for reflecting the ordered structure of an ordinal categorical variable. We have briefly described the issue of choosing scores in Section 4.3.2 but one may also peruse the pages of Bradley et al. (1962), Snell (1964), Williams and Grizzle (1972), Best and Rayner (1996), Rayner and Best (2000), Agresti (1997, 2002) and Beh and Farver (2009), for example, for further insight into this matter. These constraints imply that, from a statistical perspective, the expectation and variance of α_{iu} and β_{jv} is 0 and 1 respectively. To show this, consider the uth order row orthogonal polynomial. Then

$$
E(\alpha_{iu}) = \sum_{i=1}^{I} p_{i\bullet}\alpha_{iu}
$$
$$
= 0
$$

while

$$
\mathrm{Var}(\alpha_{iu}) = E(\alpha_{iu}^2) - [E(\alpha_{iu})]^2
$$
$$
= \sum_{i=1}^{I} p_{i\bullet}\alpha_{iu}^2 - 0
$$
$$
= 1 .
$$

The term λ_{uv} in Eq. (4.7) is the correlation between the uth order row polynomial and the vth order column polynomial such that

$$
\lambda_{uv} = \mathrm{Corr}(\alpha_{iu}, \beta_{jv})
$$
$$
= \sum_{i=1}^{I}\sum_{j=1}^{J} p_{ij} \frac{(\alpha_{iu} - E(\alpha_{iu}))}{\sqrt{\mathrm{Var}(\alpha_{iu})}} \frac{(\beta_{jv} - E(\alpha_{jv}))}{\sqrt{\mathrm{Var}(\beta_{jv})}}
$$
$$
= \sum_{i=1}^{I}\sum_{j=1}^{J} p_{ij}\alpha_{iu}\beta_{jv} \tag{4.8}
$$

and is referred to as the (u, v)th order *generalised correlation* Rayner and Beh (2009). For example, when $u = v = 1$ and natural scores are used to reflect the structure of the ordered row and column variables,

$$
\lambda_{11} = \sum_{i=1}^{I}\sum_{j=1}^{J} p_{ij}\left(\frac{i - \mu_I}{\sigma_I}\right)\left(\frac{j - \mu_J}{\sigma_J}\right) \tag{4.9}
$$

is the Pearson product moment correlation of the contingency table (Rayner and Best 1996) where

$$
\mu_I = \sum_{i=1}^{I} i p_{i\bullet} \qquad \mu_J = \sum_{j=1}^{J} j p_{\bullet j}
$$

and

$$
\sigma_I^2 = \sum_{i=1}^{I} i^2 p_{i\bullet} - \mu_I^2 \qquad \sigma_J^2 = \sum_{j=1}^{J} j^2 p_{\bullet j} - \mu_J^2 .
$$

When mid-rank scores are used instead of natural scores then λ_{11} is akin to Spearman's rank correlation (Best and Rayner 1996). In both cases, λ_{11} assesses the magnitude of the linear-by-linear association between the variables and is the maximum correlation possible between the ordered row and ordered column categories. Similarly, when $u = 2$ and $v = 1$, λ_{21} is the quadratic-by-linear association and provides one source of non-linear association when studying the association structure of the variables. Higher order non-linear sources of association can be identified by considering other values of $u > 1$ and $v > 1$ although, from a practical perspective, when $u \geq 4$ or $v \geq 4$ the interpretation of the generalised correlations becomes more difficult. In most practical situations the most dominant generalised correlations will arise when $u = 1$, 2 and/or $v = 1$, 2. This means that most of the association between two categorical variables can be captured by considering only the difference in the mean (u, $v = 1$) and variation (u, $v = 2$) of the profiles.

Another benefit of applying the BMD to the Pearson residuals when performing an ordered correspondence analysis is that $\sqrt{n}\lambda_{uv}$ is asymptotically standard normally distributed, a feature discussed in Lancaster (1953), Best and Rayner (1996) and Theorem 1 of Rayner and Best (1996). This features proves very useful for detecting statistically significant linear and non-linear sources of association.

Not only do the λ_{uv} values provide a meaningful interpretation of the association structure between the variables but the sum-of-squares of these terms gives the total inertia

$$\frac{X^2}{n} = \sum_{u=1}^{I-1}\sum_{v=1}^{J-1}\lambda_{uv}^2 \, . \tag{4.10}$$

By simply multiplying both sides of Eq. (4.10) by the sample size, n, we can determine Pearson's chi-squared statistic of the contingency table such that

$$X^2 = n\sum_{u=1}^{I-1}\sum_{v=1}^{J-1}\lambda_{uv}^2 \, . \tag{4.11}$$

More on this, and analogous, partitions of the chi-squared statistic can be found by referring to, for example, Lancaster (1953), Best and Rayner (1996), Rayner and Best (1996, 2001), Beh and Davy (1998) and Beh (1999, 2001a).

4.4.2 BMD and the Temperature Data

Recall that in Section 4.2 we found that there is a statistically significant association between the *Month* and *Temperature* variables of Table 1.2. Since these variables are ordered we decompose the Pearson residuals using BMD; see Eq. (4.7). The numerical features from this decomposition are given as follows. Table 4.3 summarises the elements of the $I - 1 = 4 - 1 = 3$ row orthogonal polynomials calculated using the scores $\mathbf{s}_I = (64.0, 75.5, 82.0, 91.0)$. Similarly, by using natural column scores $\mathbf{s}_J = (1, 2, 3, 4, 5)$, Table 4.4 gives the elements of the $J - 1 = 5 - 1 = 4$ column orthogonal polynomials. For these two tables, the first column reflects the difference in the mean of the row and column profiles and are the elements of the *linear orthogonal polynomial*. We can see that the order of the elements of this polynomial is consistent with the order imposed upon the (increasing) natural scores. For example, the first column of Table 4.3 consists of elements that range from -1.430 (for [56, 72]) to 1.376 (for [85, 97]) in a linear manner and reflect the ordered structure of the natural scores, $s_I(i) = i$ for $i = 1$, 2, 3, 4. Therefore, this row polynomial may help to reveal whether the temperature at La Guardia increases as the *Month* moves from *May* to *September*. Similarly, the second column of Table 4.3 and Table 4.4 reflects the variation of the row and column profiles, respectively. Such a quadratic structure of these polynomial elements may help to reveal whether the temperature increases, then decreases, say, as the period of study moves from *May* to *September*.

The presence, or absence, of linear or non-linear sources of association between the variables is not dependent on which orthogonal polynomial is considered. Instead, such sources of association is completely dependent on the direction and magnitude of the generalised correlations defined using Eq. (4.8). Since Table 1.2 consists of $I = 4$ rows and $J = 5$

Table 4.3 The α_{iu} values from the BMD of $\gamma_{ij} - 1$ for Table 1.2.

Temperature	Dim 1 (u = 1)	Dim 2 (u = 2)	Dim 3 (u = 3)
(56, 72]	−1.430	0.937	−0.276
(72, 79]	−0.235	−1.062	1.234
(79, 85]	0.441	−0.815	−1.428
(85, 97]	1.376	1.169	0.458

Table 4.4 The β_{jv} values from the BMD of $\gamma_{ij} - 1$ for Table 1.2.

Month	Dim 1 (v = 1)	Dim 2 (v = 2)	Dim 3 (v = 3)	Dim 4 (v = 4)
May	−1.426	1.204	−0.714	0.272
June	−0.715	−0.588	1.422	−1.089
July	−0.005	−1.184	0.016	1.581
August	0.706	−0.584	−1.397	−1.054
September	1.417	1.212	0.719	0.272

columns, there are $(4 - 1)(5 - 1) = 12$ generalised correlations and they are summarised in Table 4.5. Note that for any informative summary of the association to be made typically only the first-or second-order correlations (λ_{11}, λ_{12}, λ_{21} and λ_{22}) are required. To determine those generalised correlations that are statistically significant sources of association, recall that $\sqrt{n}\lambda_{uv}$, for $u = 1, \ldots, I - 1$ and $v = 1, \ldots, J - 1$, is an asymptotically standard normally random variable. Table 4.6 summarises these values and, in parentheses, their p-value. It shows that the linear-by-linear (λ_{11}) and linear-by-quadratic (λ_{12}) generalised correlations are statistically significant, with the latter being the most dominant. Table 4.6 also shows that there are other statistically significant sources of association between the variables but they are not as dominant as the two we just mentioned. So we shall not discuss them any further here.

The generalised correlations summarised in Table 4.5 are calculated using the mid-point row scores, $\mathbf{s}_I = (64.0, 75.5, 82.0, 91.0)$, and the natural column scores, $\mathbf{s}_J = (1, 2, 3, 4, 5)$. Therefore, $\lambda_{11} = 0.335$ is the Pearson product moment correlation between the *Month* and *Temperature* variables and implies that as the time period moves from *May* through to *September* there is an increase in the temperature. However, the magnitude of $\lambda_{12} = -0.588$ is much larger and a more dominant source of association; thus there is a strong negative linear-by-quadratic correlation between *Temperature* and *Month*. The dominance of this generalised correlation implies that as the time period moves from *May* to *September*, the temperature increases then decreases. This should be of no surprise to those in the Northern Hemisphere where such temperature variations over

Table 4.5 The generalised correlations, λ_{uv}, from the BMD of $\gamma_{ij} - 1$ for Table 1.2.

λ_{uv}	$v = 1$	$v = 2$	$v = 3$	$v = 4$
$u = 1$	0.335	−0.588	−0.022	0.003
$u = 2$	−0.078	0.222	−0.215	0.048
$u = 3$	0.049	0.113	0.056	−0.326

Table 4.6 The $\sqrt{n}\lambda_{uv}$ values and their p-value for Table 1.2.

λ_{uv}	$v = 1$	$v = 2$	$v = 3$	$v = 4$
$u = 1$	4.130	−7.244	−0.267	0.039
(p-value)	(<0.001)	(<0.001)	(0.395)	(0.484)
$u = 2$	−0.961	2.740	−2.653	0.586
(p-value)	(0.168)	(0.003)	(0.004)	(0.279)
$u = 3$	0.605	1.391	0.689	−4.014
(p-value)	(0.273)	(0.082)	(0.245)	(<0.001)

this time period are to be expected. Such a conclusion is also apparent when noting that the dominant cells in Table 1.2 follow a similar trend.

We can also see that the sum-of-squares of the elements in Table 4.5 gives the total inertia; see Eq. (4.10). That is

$$\frac{X^2}{n} = (0.335)^2 + (-0.078)^2 + \ldots + (-0.326)^2$$
$$= 0.686 .$$

Note that by multiplying this quantity by the sample size, $n = 152$, gives Pearson's chi-squared statistic of 104.3; see Eq. (4.11).

This association structure can be better viewed by constructing a graphical summary akin to the correspondence plot, or biplot, described in Chapter 2. We shall now discuss both plots in turn.

4.5 Constructing a Low-Dimensional Display

4.5.1 Standard Coordinates

Suppose we consider again Table 1.2. The elements of the row and column orthogonal polynomials that are summarised in Table 4.3 and Table 4.4 can be used as *standard coordinates* for the ordered categories of *Temperature* and *Month*. Constructing a visual summary of this association using these coordinates would imply that there exists only a linear-by-linear association between the variables. However, given the dominance of λ_{12} relative to other values of λ_{uv}, such a plot would not reveal the true nature of the association between the two ordinal variables. This is because using the orthogonal polynomials as standard coordinates does not reflect the linear and non-linear sources of association reflected in the calculation of these generalised correlations; see Table 4.5. Therefore, we do not use the orthogonal polynomials to visually summarise the association between two ordinal variables. Instead, we define *polynomial principal coordinates* to provide such a summary; in Section 4.5.2 we discuss the principal coordinates for simple ordinal correspondence analysis.

4.5.2 Principal Coordinates

To reflect the linear and non-linear sources of association between two ordinal variables of a contingency table, we define the row and column *principal coordinates* as follows. For the ith row category its principal coordinate along the vth dimension of a correspondence plot, for $v = 1, 2, \ldots, J - 1$, is

$$f_{iv} = \sum_{u=1}^{I-1} a_{iu} \lambda_{uv} .$$ (4.12)

Similarly, the principal coordinate of the jth column category along the uth dimension of a correspondence plot, for $u = 1, 2, \ldots, I - 1$, is

$$g_{ju} = \sum_{v=1}^{J-1} \beta_{jv} \lambda_{uv}$$ (4.13)

What makes these coordinates different from the principal coordinates obtained from the simple correspondence analysis of nominal variables – see Eq. (2.12) and Eq. (2.13) – is that the ordered row principal coordinates of Eq. (4.12) are plotted using at most $J - 1$ dimensions, not $S = \min(I, J) - 1$ as is the case in Chapter 2. Similarly, the ordered column principal coordinates of Eq. (4.13) are plotted in at most $I - 1$ dimensions. Therefore, a feature of this approach to correspondence analysis is that there are two explained inertia values for each dimension of the correspondence plot used to visualise the interactions between the ordered row and column categories. There is a *row inertia* that describes the differences in the row profiles in terms of their mean value, variation and higher order moments, and a *column inertia* that quantifies the differences between the column profiles. Therefore, since the row categories can be represented in a sub-space that consists of at most $J - 1$ dimensions, there are $J - 1$ explained inertia values for the row variable that can be calculated. Similarly, since the column components can be depicted in a sub-space consisting of no more than $I - 1$ dimensions, there are $I - 1$ explained inertia values for the column variable. More will be said on this feature shortly, but the interested reader may also refer to Beh (1997) and Lombardo et al. (2016a) for more information on this variant.

To gain further insight into how these principal coordinates reflect the linear, and non-linear, sources of association that may be present in the contingency table, consider

Eq. (4.12). Expanding out the summation term reveals that, for the first dimension (when $v = 1$),

$$f_{i1} = \alpha_{i1}\lambda_{11} + \alpha_{i2}\lambda_{21} + \ldots + \alpha_{i,I-1}\lambda_{I-1,1} \, .$$

Similarly, the principal coordinate of the jth column category along the first dimension (when $u = 1$) is

$$g_{j1} = \beta_{j1}\lambda_{11} + \beta_{j2}\lambda_{12} + \ldots + \beta_{j,J-1}\lambda_{1,J-1} \, .$$

These results show that the ordered categories will have an ordered arrangement along the first dimension if and only if all $\lambda_{uv} = 0$ for $u = 1, \ldots, I - 1$ and $v = 1, \ldots, J - 1$ except for λ_{11}. In this case, the configuration of the row and column principal coordinates will be ordered along the first dimension only. This feature is in stark contrast to other approaches designed to reflect the ordinal structure of categorical variables, a point we raised in Section 4.1. This is because these approaches aim to find values of a_{is} and b_{js} so that, for increasingly ordered row and column categories, $a_{1s} < a_{2s} < \ldots < a_{Is}$ and $b_{1s} < b_{2s} < \ldots < b_{Js}$, for some s such that the correlation between a_{is} and $b_{js'}$ is zero for $s \neq s'$. Therefore, these alternative approaches to ordered correspondence analysis imply that, for row and column variables consisting of categories of increasing order, $f_{1s} < f_{2s} < \ldots < f_{Is}$ and $g_{1s} < g_{2s} < \ldots < g_{Js}$. When $s = 1$, as is the case for the simple correspondence analysis of nominal variables (since λ_1 is the maximum possible correlation between the variables), this implies that the most dominant source of association between the variables will be the linear-by-linear association. This may be the case in some practical situations for the analysis of ordered variables, but it is not always going to be the case (as we have seen for Table 1.2). There may be more dominant non-linear association structures present in the data that are therefore ignored. Hence the configuration of an ordered correspondence plot using the principal coordinates defined by Eq. (4.12) and Eq. (4.13) is influenced by ALL of the generalised correlations, not just by the linear-by-linear generalised correlation. Only when all of the non-linear generalised correlations, except λ_{11}, are zero does Eq. (4.12) and Eq. (4.13) simplify to

$$f_{i1} = \alpha_{i1}\lambda_{11}$$
$$g_{j1} = \beta_{j1}\lambda_{11}$$

which are akin to the principal coordinates from the simple correspondence analysis of a nominal contingency table; see Eq. (2.12) and Eq. (2.13) for $s = 1$. Here, λ_{11} is defined by

$$\lambda_{11} = \sum_{i=1}^{I}\sum_{j=1}^{J}p_{ij}\alpha_{i1}\beta_{j1}$$

$$= \sum_{i=1}^{I}\sum_{j=1}^{J}p_{ij}\frac{(s_I(i) - \mu_I)}{\sigma_I}\frac{(s_J(j) - \mu_J)}{\sigma_J}$$

which is a more generalised form of Eq. (4.9) and is the maximum possible linear-by-linear correlation between the ordered row categories and the ordered column categories. Here μ_I and μ_J are defined by Eq. (4.2) and Eq. (4.4), respectively, while σ_I^2 and σ_J^2 are defined by Eq. (4.5) and Eq. (4.6), respectively.

We can also show that the row and column principal coordinates defined by Eq. (4.12) and Eq. (4.13), respectively, are centred at the origin. This is equivalent to the coordinate having the property that its expected value is zero. To show this for the row principal coordinate, say,

$$E(f_{iv}) = \sum_{i=1}^{I}\sum_{u=1}^{I-1}\sum_{v=1}^{J-1} p_{i\bullet} \alpha_{iu} \lambda_{uv}$$

$$= \sum_{u=1}^{I-1}\sum_{v=1}^{J-1} \lambda_{uv} \left(\sum_{i=1}^{I} p_{i\bullet} \alpha_{iu} \right)$$

$$= 0 .$$

Similarly, the variance of the row (response) principal coordinates is equivalent to the total inertia of the contingency table since

$$\text{Var}(f_{iv}) = E(f_{iv}^2) - [E(f_{iv})]^2$$

$$= \sum_{i=1}^{I}\sum_{u=1}^{I-1}\sum_{v=1}^{J-1} p_{i\bullet} f_{iv}^2 - 0$$

$$= \sum_{i=1}^{I}\sum_{u=1}^{I-1}\sum_{v=1}^{J-1} p_{i\bullet} \alpha_{iu}^2 \lambda_{uv}^2$$

$$= \sum_{u=1}^{I-1}\sum_{v=1}^{J-1} \lambda_{uv}^2 \left(\sum_{i=1}^{I} p_{i\bullet} \alpha_{iu}^2 \right)$$

$$= \sum_{u=1}^{I-1}\sum_{v=1}^{J-1} \lambda_{uv}^2$$

$$= \frac{X^2}{n}$$

using Eq. (4.10). Similar derivations can also be made to show that the column principal coordinates are centred around the origin of the correspondence plot and their variation can also be measured using the total inertia, X^2/n.

This derivation of the variation of the row principal coordinates defined by Eq. (4.12) shows that the total inertia of the contingency table can be expressed as

$$\frac{X^2}{n} = \sum_{i=1}^{I}\sum_{v=1}^{J-1} p_{i\bullet} f_{iv}^2 . \tag{4.14}$$

The total inertia may also be expressed in terms of the column principal coordinates, defined by Eq. (4.13), such that

$$\frac{X^2}{n} = \sum_{j=1}^{J}\sum_{u=1}^{I-1} p_{\bullet j} g_{ju}^2 .$$

Both these results show that, just like we described for the classical (nominal) correspondence analysis approach (see Section 2.5.2), the position of the ordered row and column principal coordinates reflects the magnitude of the total inertia, X^2/n. When the profile coordinates are close to the origin then their row, or column, category does not contribute greatly to the association between the ordered variables since they only make a small contribution to the total inertia. On the other hand, principal coordinates that are positioned far from the origin display those categories that do make such a contribution. In fact, Beh (1997) showed that the difference between the ith row profile and the i'th row profile when performing ordinal correspondence analysis is reflected by the squared Euclidean distance of their position in the optimal correspondence plot (consisting of $I - 1$ dimensions) since

$$d_I^2(i,\ i') = \sum_{j=1}^{J} \frac{1}{p_{\bullet j}} \left(\frac{p_{ij}}{p_{i\bullet}} - \frac{p_{i'j}}{p_{i'\bullet}} \right)^2$$

$$= \sum_{j=1}^{J} \frac{1}{p_{\bullet j}} \left[\left(\frac{p_{ij}}{p_{i\bullet}} - p_{\bullet j} \right) - \left(\frac{p_{i'j}}{p_{i'\bullet}} - p_{\bullet j} \right) \right]^2$$

$$= \sum_{v=1}^{I-1} (f_{iv} - f_{i'v})^2 \ .$$

Hence, the property of distributional equivalence is preserved in this variant of correspondence analysis. The difference between the ith row profile, say, and the mean profile is also reflected by the squared Euclidean distance of its point from the origin since

$$d_I^2(i,\ 0) = \sum_{j=1}^{J} \frac{1}{p_{\bullet j}} \left(\frac{p_{ij}}{p_{i\bullet}} - p_{\bullet j} \right)^2$$

$$= \sum_{v=1}^{J-1} f_{iv}^2 \ .$$

Thus, since the total inertia can be expressed in terms of the row principal coordinates by Eq. (4.14), the total inertia of the doubly ordered contingency table can be expressed as the weighted sum-of-squares of these distances such that

$$\frac{X^2}{n} = \sum_{i=1}^{I} p_{i\bullet} d_I^2(i,\ 0) \ .$$

Similar results can also be obtained using the column principal coordinated defined by Eq. (4.13).

4.5.3 Practicalities of the Ordered Principal Coordinates

Recall that using the principal coordinates defined by Eq. (4.12) and Eq. (4.13) leads to two explained inertia values along each dimension of the correspondence plot; each dimension reflecting a source of difference between the row profiles and the column profiles. As a result, Beh and Lombardo (2014, p. 230) advocate that if these coordinates are to be used to visualise the mean and variation differences in the profiles then separate plots be constructed; one for the row variable and another for the column variable.

While it is advisable that a joint display not be produced using the principal coordinates defined by Eq. (4.12) and Eq. (4.13), they can be used for constructing an isometric biplot. We shall now turn our attention to this type of visual summary.

4.6 The Biplot Display

4.6.1 Definition

A more appropriate way to visualise the association between the ordered row and column categories of a two-way contingency table using doubly ordered correspondence analysis is to produce a biplot. Like we discussed in Section 2.7 for the correspondence analysis of nominal variables, for the analysis of two ordered variables the two types of biplots we will discuss here are the *ordered row isometric biplot* and the *ordered column isometric biplot*. These two biplots can be interpreted in the same way as the biplots described in Chapter 2. We now discuss them in turn.

4.6.2 Ordered Column Isometric Biplot

For an ordered column isometric biplot, the biplot coordinate for the ith row category and the jth column category along the uth dimension of the biplot is

$$\tilde{f}_{iu} = \alpha_{iu} \tag{4.15}$$

$$\tilde{g}_{ju} = \sum_{v=1}^{J-1} \beta_{jv} \lambda_{uv} \tag{4.16}$$

for $u = 1, 2, \dots, I-1$. Such a biplot is therefore constructed using the standard coordinates (that is, the elements of the uth order row orthogonal polynomial, for $u = 1, \dots, I-1$) of the row categories and the principal coordinates of the column categories. By using Eq. (4.15) and Eq. (4.16) to visually depict the association between the ordered variables, the (i, j)th Pearson residual can be expressed by

$$\gamma_{ij} - 1 = \sum_{u=1}^{I-1} \tilde{f}_{iu} \tilde{g}_{ju}$$

and is the inner product of the ith row standard coordinate and the jth column principal coordinate. Thus, if this residual is large (so that the position of the jth column principal coordinate is in close proximity to the projection of the ith row standard coordinate from the origin) then there is a strong interaction between this row/column pair. Similarly, if the residual is small (indicating a lack of interaction between this pair) then the column principal coordinate will be positioned a long way from the projection of the row.

4.6.3 Ordered Row Isometric Biplot

The second kind of biplot we consider here is the ordered row isometric biplot. For this type of visual summary, the biplot coordinate of the ith row category and the jth column category is defined so that

$$\tilde{f}_{iv} = \sum_{u=1}^{I-1} \alpha_{iu} \lambda_{uv}$$

$$\tilde{g}_{jv} = \beta_{jv}$$

for $v = 1, 2, \ldots, J - 1$. In this case, the row categories are depicted using their principal coordinate and the column categories are depicted using the elements of the vth order column orthogonal polynomial (which is analogous to their standard coordinate). For this definition of coordinates, the (i, j)th Pearson residual can be expressed by

$$\gamma_{ij} - 1 = \sum_{v=1}^{J-1} \tilde{f}_{iv} \tilde{g}_{jv}$$

and is the inner product of the ith row principal coordinate and the jth column standard coordinate. Therefore, if this residual is large (so that the position of the ith row principal coordinate is in close proximity to the projection of the jth column standard coordinate from the origin) then there is a strong interaction between this row/column pair. Similarly, if the residual is small (indicating a lack of interaction between these two categories) then the row principal coordinate will be positioned a long way from the projection of the column.

4.6.4 Ordered Isometric Biplots for the Temperature Data

Consider again the temperature data summarised in Table 1.2. Figure 4.4 gives the row isometric biplot when the correspondence analysis approach described in this chapter is applied. Here, the categories of the *Month* variable are depicted as projections from the origin to their standard coordinate (column orthogonal polynomials) while the *Temperature* categories are depicted as points defined by their principal coordinates. To help show the quadratic pattern of the standard coordinates along the second dimension, a light dashed line joins the row points.

Figure 4.4 shows that the temperature in the month of *May* is more likely to be between 56 and 72 degrees Fahrenheit than any of the temperature ranges in Table 1.2; this can be seen from the close proximity of (56, 72] from the projection of *May*. During the months of *July* and *August*, the temperature at La Guardia airport ranges between 79 degrees and 97, with *July* experiencing temperatures in the lower half of this range and *August* experience hotter temperatures. Since (72, 79] lies very close to the origin, this is deemed to be the average temperature range at the airport at the time the data was collected; in fact, from the airquality dataset, the mean temperature recorded was 77.9 degrees Fahrenheit.

Suppose we now turn our attention to the explained inertia values for the *Temperature* variable that is visually depicted in Figure 4.4. The row explained inertia along the first dimension of Figure 4.4 is $\sum_{u=1}^{J-1} \lambda_{u1}^2 = 0.121$ and contributes to $100 \times 0.121/0.686 = 17.59\%$ of the total inertia. That is, the first dimension of this row isometric biplot displays only 17.59% of the inertia between the ordered *Temperature* categories and highlights differences in the mean values of the row profiles. Similarly, the row explained inertia along the second dimension is $\sum_{u=1}^{I-1} \lambda_{u2}^2 = 0.407$ and contributes to $100 \times 0.407/0.686 = 59.36\%$ of the total inertia. This means that more than half of the differences that exist between the profiles of the *Temperature* categories can be accounted for by the difference in their variation. The dominance of this second explained inertia is clear by observing the configuration of points for the *Temperature* categories in Figure 4.4 which are strongly dominated by their position along the second (variation) dimension rather than along the first (mean) dimension. Recall that this was also the case when the classical approach to correspondence analysis is applied; see Figure 4.1. In fact, not only does Figure 4.4 show that the *Temperature* range

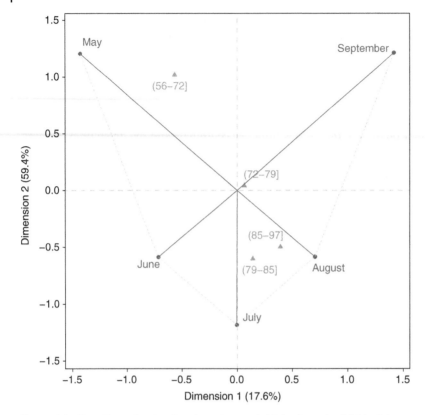

Figure 4.4 Two-dimensional ordered row isometric biplot from the DOCA of the temperature data summarised in Table 1.2.

(56, 72] has a strong interaction with the *Month* of *May*, it also shows that the mean and variation of this rows profile is more extreme than the mean and variation of all of the other row categories. This is evident by observing that its position in the row isometric biplot is the furthest row point from the origin. This conclusion is consistent with our finding that $\mu_I(1) = -0.796$ is furthest away from zero than any other row mean value and also exhibits the greatest variation of the row profiles with $\sigma_I^2(1) = 13.785$. The *Temperature* range with a profile that is closest to the overall mean row profile is (72, 79]. Recall that its profile has a mean value of $\mu_I(2) = 0.091$. These findings can be made by observing that its position in Figure 4.4 is very close to the origin. Therefore, this profile is the least likely of the *Temperature* ranges to contribute to the association that exists between the two variables of Table 1.2. One must be careful to not draw too strict a parallel between the values derived in Section 4.3.5 and the inertia value calculated using orthogonal polynomials. Instead, the values of $\mu_I(i)$ and $\sigma_I^2(i)$, when studying the rows, may only be considered as indicative of the type of conclusions one may draw from visually summarising the association between ordered row and column variables using an isometric biplot.

With a mean and variation row explained inertia values contributing to 17.59% and 59.36% of the total association, respectively, Figure 4.4 visually summarises 76.95% of the association between the *Month* and *Temperature* variables of Table 1.2. The remaining

23.05% reflects higher order sources of variation. Since, for the row categories, there are $J - 1 = 5 - 1 = 4$ inertia values, the third and fourth inertia values account for 7.27% and 15.78%, respectively, and describe the differences between the row profiles in terms of their skewness and kurtosis. Thus, we can see here that the difference in the skewness coefficients of the row profiles is as dominant as the differences in their mean values.

We now turn our attention to the *Month* variable and the quality of column isometric plot given by Figure 4.5. The column explained inertia along its first and second dimensions is $\sum_{v=1}^{J-1} \lambda_{1v}^2 = 0.458$ and $\sum_{v=1}^{J-1} \lambda_{2v}^2 = 0.104$, respectively. Therefore, the first dimension of Figure 4.5 visually describes $100 \times 0.458/0.686 = 66.73\%$ of the differences that exist in the *Month* (column) profiles while the second dimension reflects only $100 \times 0.104/0.686 = 15.16\%$ of the total inertia. These quantities show that differences between the profile of the *Month* categories is dominated more by their differences in their mean values than in their variation. This is evident by observing Figure 4.5 where the configuration of column points dominates along the first dimension. Therefore, Figure 4.5 visually accounts for 81.89% of the total inertia. The remaining 18.11% of the inertia reflects higher order sources of variation. Since, for the column categories, there are $I - 1 = 4 - 1 = 3$ explained inertia values, the third explained inertia is 0.124 and accounts for 18.11% of the total inertia. This inertia

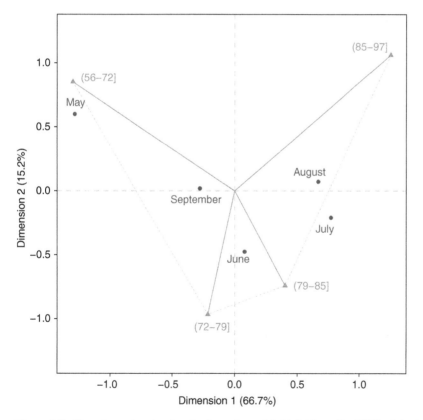

Figure 4.5 Two-dimensional ordered column isometric biplot from the DOCA of the temperature data summarised in Table 1.2 with beta scaling ($\beta = 1.1$).

value reflects the skewness differences in the column profiles. Therefore, we can see here that the differences in the skewness coefficient of the column profiles is more dominant than the differences in their variation.

From Figure 4.5, we can draw very similar conclusions about the association between the two variables to those we made for Figure 4.4. Although for the column isometric biplot, these conclusions can be reached by observing the proximity of the *Temperature* categories from the projections of the *Month* categories from the origin. For example, both Figure 4.4 and Figure 4.5 show the month of *May* and the *Temperature* range (56, 72] have a strong interaction with each other. In fact, Figure 4.5 shows that the column profile with the greatest variation and the most extreme mean value is for the *Month* of *May*; of the five months in which the study was undertaken, the position of *May* is the furthest from the origin. While, the mean and variation values calculated in Section 4.3.5 are only indicative, we saw that $\mu_J(1) = -11.243$ and $\sigma_J^2(1) = 41041.864$ are the most extreme measures of mean and variation. Note also that in Figure 4.5, *May* and *September* are located on the left side of the origin suggesting that their *Temperature* is less than the average during the period of the study; note that $\mu_J(1)$ and $\mu_J(5)$ are both negative quantities. On the other hand, the *Months* of *July* and *August* experience the hottest temperatures during the five month period.

4.7 Final Comments

In this chapter we have described one approach to simple correspondence analysis that accommodates the structure of ordinal categorical variables. This approach is based on using the orthogonal polynomials described in Emerson (1968) and has many practical advantages. Firstly, they are very simple to calculate and provide a detailed analysis of any sources of linear and non-linear association that may exist between the variables. We discussed that such quantities are asymptotically normally distributed which makes them ideal for detecting those sources that are statistically significant. The approach described in this chapter also lends itself nicely to the construction of a row, or column, isometric biplot. Unlike other approaches that have been presented in the past, the correspondence analysis approach described in this chapter does not require that the order of the categories be preserved in the correspondence plot. This is because we are not just focusing on the presence (or not) of a linear-by-linear association between the variables but we are also incorporating the non-linear sources of association we just spoke of. Despite this, work is ongoing to generalise the construction of confidence ellipses presented in Section 2.9. However, one may construct confidence circles that are of the same type described by Lebart et al. (1984) for nominal variables; see, for example, Beh (2001a) who showed that their circles can be constructed for a doubly ordered contingency table.

A feature of the association structure that we have yet to discuss is where there exists an asymmetric association between a row variable and a column variable of a two-way contingency table where at least one of these variables consists of ordered categories. Therefore, the next step in our discussion is to explore the development of non-symmetrical correspondence analysis for ordered categorical variables. Such an approach was first described in detail by Lombardo et al. (2011) and we provide an overview of it in the next chapter for asymmetrically associated categorical data.

5

Ordered Non-symmetrical Correspondence Analysis

5.1 Introduction

Back in Chapter 3 we described how non-symmetrical correspondence analysis can be applied to a two-way contingency table formed from the cross-classification of two nominal variables. There we treated the column variable as the predictor variable and the row variable as the response variable; we will continue with an examination of this asymmetric association structure for our discussions in this chapter. Of course, there are times when one, or both, variables consist of ordered categories and this structure needs to be incorporated into the analysis. Recall that we did just that for two symmetrically associated variables in Chapter 4. In this chapter we shall provide an overview of how non-symmetrical correspondence analysis can be performed on a *singly ordered contingency table*; that is, a contingency table consisting of one nominal (row/response) variable and one (column/predictor) ordinal variable. As a result, this variant of correspondence analysis is referred to as *singly ordered non-symmetrical correspondence analysis*.

The approach that we describe for studying this association is akin (but not identical) in structure to the doubly ordered classical approach we described in Chapter 4. For a singly ordered analysis of two asymmetrically associated variables we use a *hybrid decomposition* instead of a bivariate moment decomposition (BMD) or generalised singular value decomposition (GSVD). Since we are treating the row variable as consisting of nominal, and response, categories and the column variable as consisting of ordered, and predictor, categories, the *hybrid* nature of the decomposition is formed by jointly incorporating the numerical row features from the singular value decomposition of the centred column profiles and the numerical column features from the bivariate moment decomposition of these profiles. So, for the purposes of describing singly ordered non-symmetrical correspondence analysis, we shall be applying our hybrid decomposition to the centred column profiles rather than using a GSVD like we did in Chapter 3; see Eq. (3.2).

One may also examine the non-symmetrical correspondence analysis of two ordered categorical variables, as Lombardo et al. (2007) and Beh and Lombardo (2014, Section 7.3) did. For such an approach, rather than applying a GSVD to the centred column profiles

An Introduction to Correspondence Analysis, First Edition. Eric J. Beh and Rosaria Lombardo.
© 2021 John Wiley & Sons Ltd. Published 2021 by John Wiley & Sons Ltd.

as we did for the study of two asymmetrically associated variables (see Eq. (3.2)), a BMD may instead be used. Since we discussed the application of this method of decomposition in Chapter 4 our focus will instead be on the hybrid decomposition of the centred column profile and invite the interested reader to peruse the pages of Lombardo et al. (2011) for more information on this variant.

Our discussion of singly ordered non-symmetrical correspondence analysis in this chapter will be centred on the analysis of the shoplifting data summarised in Table 1.3. In doing so, we will be treating the *Age* (column) variable, that consists of ordered categories, as the predictor variable and the *Item* (row) variable as the nominal and response variable. Therefore, the focus of our analysis will be to explore how the age of the perpetrator helps to predict the type of item they are likely to steal.

5.2 The Goodman–Kruskal tau Index Revisited

Recall, that for non-symmetrical correspondence analysis, the key measure of association involves the Goodman–Kruskal tau index

$$\tau = \frac{\tau_{num}}{1 - \sum_{i=1}^{I} p_{i\bullet}^2}$$

where

$$\tau_{num} = \sum_{i=1}^{I} \sum_{j=1}^{J} p_{\bullet j} \left(\frac{p_{ij}}{p_{\bullet j}} - p_{i\bullet} \right)^2 .$$

A definition of this statistic was also given by Eq. (1.7). When performing non-symmetrical correspondence analysis on a two-way contingency table, the key measure of association is the total inertia defined by τ_{num} which, as was shown in Chapter 3, can be partitioned into weighted and centred column profiles such that

$$\tau_{num} = \sum_{i=1}^{I} \sum_{j=1}^{J} \tilde{c}_{i|j}^2 .$$

where

$$\tilde{\mathbf{c}}_j = (\tilde{c}_{1|j}, \ \dots \ , \tilde{c}_{I|j})$$
$$= \left(\sqrt{p_{\bullet j}} \left(\frac{p_{1j}}{p_{\bullet j}} - p_{1\bullet} \right), \ \dots \ , \sqrt{p_{\bullet j}} \left(\frac{p_{Ij}}{p_{\bullet j}} - p_{I\bullet} \right) \right) .$$

See Eq. (3.1) for a definition of these profiles. A visual depiction of these profiles can be made to compare the relative distribution (weighted by $\sqrt{p_{\bullet j}}$) of each of the columns. For Table 1.3, such a visual depiction is given by Figure 5.1. It shows that, due to the positive sign and magnitude of their $\tilde{c}_{i|j}$ value, the strongest asymmetric associations exist as follows: males aged less than 12 years of age are most likely to steal *toys*, males aged between 21 and 29 years of age (with a label of *25*) are most likely to steal *clothing* and males aged 12 to 14 years of age (with *13* as its label) have the greatest propensity towards stealing *stationary*. Such conclusions show that it is these three cells of Table 1.3 that have frequencies

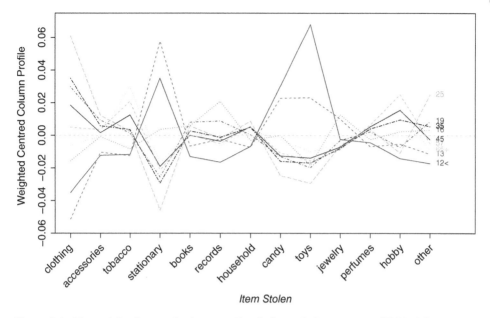

Figure 5.1 The weighted centred column profiles, $\tilde{c}_j$, for each *Age* category of Table 1.3.

that are greater than what is expected if there were no association between the *Age* and *Item* variables. Figure 5.1 also shows the following three very strong negative asymmetric associations: males aged 12 to 14 years of age are not at all likely to steal *clothing*, those aged 21 to 29 years of age are very unlikely to steal *stationary* while males aged less than 12 years of age are highly unlikely to steal *clothing*. These findings highlight the three cells of Table 1.3 whose observed count is much lower than what is expected if there were independence between these two variables. Figure 3.1 also shows that the least "variable", or flattest, profile is that of the *Age* category 16. Therefore, this age group has the most "average" propensity of stealing items with no single item being the most, or least, attractive to steal. The weighted centred column profiles, defined by Eq. (3.1) are also summarised in the columns of Table 5.1.

To determine whether the *Age* variable is a statistically significant predictor of the *Item* that is stolen, we calculate the *C*-statistic. To do so, we first determine the Goodman–Kruskal tau index of Table 1.3; to three decimal places, the index is $\tau = 0.041$ where $\tau_{num} = 0.038$. Therefore, the *C*-statistic is

$$C = (n - 1)(I - 1)\tau$$
$$= (20819 - 1)(13 - 1) \times 0.041$$
$$= 10331.51 \ .$$

Recall that the *C*-statistic is a chi-squared random variable; see Eq. (1.8). Then, at the 5% level of significance and with $(I - 1)(J - 1) = (13 - 1)(9 - 1) = 96$ degrees of freedom, this statistic has a p-value that is less than 0.001 showing that the age of the perpetrator is a statistically significant predictor of the items that are stolen.

Table 5.1 The weighted centred column profiles, $\tilde{c}_j$, of Table 1.3.

Item	Age								
	< 12	13	16	19	25	35	45	57	65+
clothing	−0.035	−0.051	−0.016	0.030	0.061	0.035	0.019	0.005	−0.006
accessories	−0.012	−0.010	−0.001	0.009	0.012	0.006	0.002	0.004	−0.001
tobacco	−0.012	−0.012	−0.008	0.002	0.003	0.004	0.013	0.021	0.030
stationary	0.035	0.058	0.004	−0.025	−0.046	−0.029	−0.019	−0.021	−0.017
books	−0.013	−0.007	0.006	0.008	0.006	0.003	0.000	0.000	0.003
records	−0.016	−0.002	0.021	0.009	−0.002	−0.001	−0.003	−0.005	−0.006
household	−0.007	−0.007	−0.003	0.000	0.009	0.005	0.005	0.005	0.003
candy	0.030	0.023	0.000	−0.013	−0.025	−0.016	−0.013	−0.011	−0.003
toys	0.068	0.023	−0.018	−0.020	−0.029	−0.017	−0.014	−0.016	−0.010
jewelry	−0.002	0.010	0.013	−0.003	−0.006	−0.008	−0.007	−0.009	−0.005
perfumes	−0.004	−0.007	−0.003	0.002	0.004	0.004	0.005	0.006	0.005
hobby	−0.014	−0.005	0.002	−0.007	−0.011	0.010	0.015	0.025	0.016
other	−0.017	−0.011	0.003	0.009	0.025	0.005	−0.003	−0.005	−0.009

5.3 Decomposing $\pi_{i|j}$ for Ordinal and Nominal Variables

5.3.1 The Hybrid Decomposition of $\pi_{i|j}$

Consider again our two-way contingency table **N** where the column (predictor) variable consists of ordered categories while the row (response) variable consists of nominal categories. In this case, the structure of the rows can be reflected by using the set of nominal row scores and the columns can be reflected using ordered column scores. However, the choice of nominal scores is virtually endless and so a more strategic approach is to treat the nominal row structure in the same manner as we did in Chapters 2 and 3 through the method of decomposition; we shall speak more about this shortly. For the ordered column categories of Table 1.3, the column scores that we shall use are $\mathbf{s}_J = (5.5, 13.0, 16.0, 19.0, 25.0, 34.5, 44.5, 57.0, 72.0)$. These scores represent the mid-point of the nine age groups that define the *Age* variable of Table 1.3 and serve as the basis vector for determining the column orthogonal polynomials; see Section 4.4.1.

For the analysis of the asymmetric association between the two variables of a two-way contingency table **N**, we can visually describe this association using *singly ordered non-symmetrical correspondence analysis*. As we described in Chapter 3, the non-symmetrical correspondence analysis of two nominal categorical variables involves the GSVD of $\pi_{i|j}$, defined by Eq. (3.2). However, when one of the variables consists of ordered categories (as we have here for the columns of Table 1.3) the method of decomposition we use is referred to as *hybrid decomposition* (HD). This decomposition of $\pi_{i|j}$ takes the form

$$\pi_{i|j} = \frac{p_{ij}}{p_{\bullet j}} - p_{i\bullet} = \sum_{s=1}^{S} \sum_{v=1}^{J-1} a_{is} \lambda_{s(v)} \beta_{jv} . \tag{5.1}$$

Here a_{is} is the ith element of the sth left singular vector of the matrix of $\pi_{i|j}$ values and is constrained by

$$\sum_{i=1}^{I} a_{is} = 0, \qquad \sum_{i=1}^{I} a_{is}^2 = 1 . \tag{5.2}$$

Similarly, β_{jv} is the jth element of the vth order column orthogonal polynomial. Like those we described in Chapter 4 for the *doubly ordered correspondence analysis* of a two-way contingency table, these polynomials can be derived using the simple recurrence formulae of Emerson (1968). These polynomials are constrained so that

$$\sum_{j=1}^{J} p_{\bullet j} \beta_{jv} = 0, \qquad \sum_{j=1}^{J} p_{\bullet j} \beta_{jv}^2 = 1 . \tag{5.3}$$

The constraints of Eq. (5.2) and Eq. (5.3) imply that the expectation and variance of the a_{is} and β_{jv} values is zero and one, respectively. The term $\lambda_{s(v)}$ in Eq. (5.1) is the correlation between a_{is} and β_{jv} such that

$$\lambda_{s(v)} = \text{Cov}(a_{is}, \beta_{jv})$$

$$= \sum_{i=1}^{I} \sum_{j=1}^{J} \frac{(a_{is} - \text{E}(a_{is}))}{\sqrt{Var(a_{is})}} \frac{(\beta_{jv} - \text{E}(\beta_{jv}))}{\sqrt{Var(\beta_{jv})}}$$

$$= \sum_{i=1}^{I} \sum_{j=1}^{J} p_{ij} a_{is} \beta_{jv} \tag{5.4}$$

and is referred to as the (s, v)th *generalised asymmetric correlation* between the nominal row (response) variable and the ordered column (predictor) variable. It may also be expressed in terms of $\pi_{i|j}$ by

$$\lambda_{s(v)} = \sum_{i=1}^{I} \sum_{j=1}^{J} p_{\bullet j} \pi_{i|j} a_{is} \beta_{jv} .$$

Beh (2001b) showed that, for *symmetrically* associated row (nominal) and column (ordinal) variables, the sum-of-squares of these terms gives the total inertia of simple correspondence analysis

$$\frac{X^2}{n} = \sum_{s=1}^{S} \sum_{v=1}^{J-1} \lambda_{s(v)}^2$$

where $S = \min(I, J) - 1$. For *asymmetrically* associated row (nominal/response) and column (ordinal/predictor) variables, the sum-of-squares of terms defined by Eq. (5.4) gives the total inertia for non-symmetrical correspondence analysis

$$\tau_{num} = \sum_{i=1}^{I}\sum_{j=1}^{J} p_{\bullet j}\pi_{i|j}^{2}$$

$$= \sum_{s=1}^{S}\sum_{v=1}^{J-1} \lambda_{s(v)}^{2} \ . \tag{5.5}$$

Therefore, by equating Eq. (5.5) with Eq. (3.5) gives

$$\lambda_{s}^{2} = \sum_{v=1}^{J-1} \lambda_{s(v)}^{2} \ .$$

This means that the weight given to the sth dimension of the correspondence plot constructed from a non-symmetrical correspondence analysis can be partitioned into sources of association that reflect mean and variation differences in the profiles. For example, for the first dimension of a non-symmetrical correspondence plot,

$$\lambda_{1}^{2} = \lambda_{1(1)}^{2} + \lambda_{1(2)}^{2} + \ldots + \lambda_{1(J-1)}^{2}$$

where $\lambda_{1(1)}^{2}$ determines that part of λ_{1}^{2} that describes differences in the ordered column profiles in terms of their mean values, while $\lambda_{1(2)}^{2}$ quantifies the differences in the variation of the profiles.

Rather than describing the asymmetrical association between the two categorical variables in terms of τ_{num} suppose we consider its C-statistic. It can be expressed as the sum-of-squares of the weighted $\lambda_{s(v)}^{2}$ values such that

$$C = \sum_{s=1}^{S}\sum_{v=1}^{J-1} \tilde{\lambda}_{s(v)}^{2} \ .$$

where

$$\tilde{\lambda}_{s(v)} = \lambda_{s(v)} \sqrt{\frac{(n-1)(I-1)}{1 - \sum_{i=1}^{I} p_{i\bullet}^{2}}} \tag{5.6}$$

and are asymptotically standard normally distributed. Thus, the statistical significance of each $\tilde{\lambda}_{s(v)}$ can be easily assessed. This also means that the C-statistic can be partitioned into components that reflect the various sources of association that exist in the column categories. For example, the differences in the mean of the ordered column (predictor) profiles can be assessed by calculating the *mean component*

$$\tilde{\lambda}_{\bullet(1)}^{2} = \sum_{s=1}^{S} \tilde{\lambda}_{s(1)}^{2}$$

while the differences in the variation of the column profiles can be quantified by the *variation component*

$$\tilde{\lambda}_{\bullet(2)}^{2} = \sum_{s=1}^{S} \tilde{\lambda}_{s(2)}^{2} \ .$$

In general, we can assess the statistical significance of the vth order column component by

$$\tilde{\lambda}_{\bullet(v)}^{2} = \sum_{s=1}^{S} \tilde{\lambda}_{s(v)}^{2}$$

and compare it with the $1 - \alpha$ percentile of the chi-squared distribution with $J - 1$ degrees of freedom. Note that, when performing this approach to singly ordered non-symmetrical correspondence analysis, the C-statistic can be expressed in terms of the sum-of-squares of the $\tilde{\lambda}^2_{\bullet(v)}$ values. That is,

$$C = \sum_{v=1}^{J-1} \tilde{\lambda}^2_{\bullet(v)} .$$

For more on the use of Emerson's (1968) simple recurrence formulae to generate orthogonal polynomials in non-symmetrical correspondence analysis, we direct the interested reader to Beh and Lombardo (2014, Chapter 7). One may also reflect back to Chapter 4, or the papers of Beh (1997, 1998, 2004a), Beh et al. (2007), Lombardo et al. (2007), Lombardo and Beh (2010), Lombardo and Meulman (2010), Lombardo et al. (2011), Simonetti et al. (2011), Lombardo and Beh (2016) and Lombardo et al. (2016a) and Lombardo et al. (2016a) for additional insights into use of orthogonal polynomials for performing a correspondence analysis to a two-way contingency table formed from cross-classifying ordinal categorical variables.

5.3.2 Hybrid Decomposition and the Shoplifting Data

Since we have established in Section 5.2 that, for Table 1.3, the (ordered) *Age* of the perpetrator is a statistically significant predictor of the (nominal) *Item* that is stolen, we shall examine this asymmetric association visually using singly ordered non-symmetrical correspondence analysis. Before we move on to the construction of the visual display, we first identify the numerical terms that come from the hybrid decomposition of the π_{ij} terms of Table 1.3. To reflect the ordered structure of the *Age* variable, its orthogonal polynomials are constructed using the mid-point of the nine age groups, s_J, previously defined. By using these scores, the first three column orthogonal polynomials are summarised in the columns of Table 5.2 while the columns of Table 5.3 summarise the first three row (left) singular vectors from the GSVD of the π_{ij} values of Table 1.3.

Table 5.2 The β_{jv} values of Table 1.3.

Age	Dim 1 ($v = 1$)	Dim 2 ($v = 2$)	Dim 3 ($v = 3$)
< 12	−1.063	1.433	−1.331
13	−0.552	0.184	0.338
16	−0.348	−0.209	0.606
19	−0.144	−0.539	0.689
25	0.265	−1.017	0.417
35	0.912	−1.271	−0.722
45	1.593	−0.874	−1.968
57	2.444	0.581	−1.984
65+	3.465	3.733	3.027

Table 5.3 The a_{is} values of Table 1.3.

Item	Dim 1	Dim 2	Dim 3
clothing	−0.559	0.463	−0.221
accessories	−0.131	0.051	0.039
tobacco	−0.139	−0.497	−0.303
stationary	0.545	0.041	0.274
books	0.092	−0.027	0.152
records	−0.047	0.003	0.471
household	−0.088	−0.038	−0.073
candy	0.300	0.041	−0.060
toys	0.447	0.324	−0.604
jewelry	0.080	0.032	0.322
perfumes	−0.069	−0.086	−0.091
hobby	−0.083	−0.583	−0.107
other	−0.165	0.276	0.201

Table 5.4 The generalised asymmetric correlations, $\lambda_{s(v)}$, of Table 1.3.

s/v	1	2	3	4	5	6	7	8	% Total
1	−0.115	0.108	−0.023	−0.035	0.047	−0.030	0.013	0.012	79.9%
2	−0.041	−0.026	−0.004	0.028	−0.020	0.002	0.008	−0.009	10.0%
3	−0.017	−0.024	0.032	−0.029	0.010	0.003	−0.011	0.012	8.3%
4	−0.001	0.008	0.006	0.004	−0.008	0.009	−0.010	0.001	1.0%
5	−0.001	−0.004	−0.010	−0.001	0.002	0.001	−0.011	0.003	0.7%
6	<0.001	<0.001	<0.001	<0.001	−0.003	−0.002	0.002	0.007	0.2%
7	<0.001	<0.001	−0.001	−0.002	−0.003	−0.001	<0.001	−0.001	<0.1%
8	<0.001	<0.001	−0.001	−0.001	<0.001	0.002	0.001	<0.001	<0.1%
$\lambda_{\bullet(v)}$	0.015	0.013	0.002	0.003	0.001	0.001	<0.001	<0.001	0.0375
%	40.4%	34.6%	4.5%	7.6%	7.4%	2.7%	1.5%	1.1%	100.0%

The set of generalised asymmetric correlations of Table 1.3, calculated using Eq. (5.4), are summarised in Table 5.4. Since $J − 1 = 9 − 1 = 8$ and $S = \min(13, 9) − 1 = 8$ the matrix of generalised asymmetric correlation values is of size 8×8.

Table 5.4 shows that the most dominant generalised asymmetric correlation term is $\lambda_{1(1)} = −0.115$. This value shows that there is a negative mean component along the first dimension ($v = 1$) of the correspondence plot. That is, for the first dimension of the correspondence plot obtained from performing a non-symmetrical correspondence analysis on Table 1.3, the most dominant source of difference in the column profiles is in the mean values of the *Age* of perpetrators. Thus, the age of most of the perpetrators is younger than the mean. Another dominant feature along this dimension are the differences in the profiles due to their variation; this is evident by observing that $\lambda_{1(2)} = 0.108$. The relatively large values of $\lambda_{1(1)}$, compared to the values of $\lambda_{s(v)}$, for $v \neq 1$, show the dominance of the first dimension. In fact, this dimension reflects 40.4% of the asymmetric association that exists between the two categories, while the second dimension reflects 34.6%. Therefore, a two-dimensional visualisation of the association will provide a visual summary of about 75% of the association between the *Age* and *Item* variables. The dominance of $\lambda_{1(1)}$ and $\lambda_{1(2)}$ along the first dimension also means that the mean and variation components in the optimal non-symmetrical correspondence biplot are the most dominant sources of variation amongst the column profiles; we shall describe the construction of such a visual summary in Section 5.4. The differences in the mean values of the *Age* profiles account for 40.4% of the variation that exists between them, while the difference in their variation accounts for a further 34.6% of the variation. Higher order sources of association account for the remaining 15% of the variation and are all relatively small.

We also note that the sum-of-squares of the $\lambda_{s(v)}$ values summarised in Table 5.4 give the numerator of the Goodman–Kruskal tau index so that

$$\tau_{num} = (-0.115)^2 + (-0.041)^2 + \ldots + (0.000)^2$$
$$= 0.0358$$

and is the total inertia from performing a singly-ordered non-symmetrical correspondence analysis on Table 1.3.

We can assess the statistical significance of the $\lambda_{s(v)}$ terms using Eq. (5.6). For $\lambda_{1(1)}$, it can be rescaled to be a standard normally distributed random variable so that

$$\tilde{\lambda}_{1(1)} = \lambda_{1(1)} \sqrt{\frac{(n-1)(I-1)}{1 - \sum_{i=1}^{I} p_{i\bullet}^2}}$$

$$= -0.115 \sqrt{\frac{(20819 - 1)(9 - 1)}{0.908}}$$

$$= -49.262 .$$

Therefore, since $\tilde{\lambda}_{1(1)}$ is asymptotically a normally distributed random variable, the mean differences between the profile of each *Item* category is a statistically significant contributor to the asymmetric association along the first dimension of the correspondence plot. In fact, this term contributes to

$$100 \times \frac{\tilde{\lambda}_{1(1)}^2}{C} = 100 \times \frac{(-49.262)^2}{10331.51}$$
$$= 23.52\%$$

of the asymmetric association that exists between the *Age* and *Item* variables.

5.4 Constructing a Low-Dimensional Display

5.4.1 Standard Coordinates

To visually summarise the asymmetric association between the *Age* and *Item* variables of Table 1.3, we can perform a singly ordered non-symmetrical correspondence analysis on the data. Such a visual summary could be obtained using the standard coordinates for the two variables. That is, the left (row) singular values given in Table 5.3 could be used as the standard coordinates for the nominal row (response) categories while the column orthogonal polynomials (see Table 5.2) could be used as the standard coordinates for the ordered column (predictor) categories. However, using these coordinates to obtain a visualisation does not reflect the structure of the association captured in the $\lambda_{s(v)}$ values summarised in Table 5.4. Therefore, we shall not be using standard coordinates to visualise the asymmetric association between the *Age* and *Item* variables. Note that a similar argument was given in Chapters 2 and 3 where we pointed out the problems in using singular vectors to visualise the association between two nominal categorical variables; see our discussion in Section 2.5.1 for simple correspondence analysis and Section 3.4.1 for a similar discussion made for non-symmetrical correspondence analysis.

5.4.2 Principal Coordinates

An alternative to standard coordinates is to instead define principal coordinates for the row and column categories. For the ith row (nominal/response) category along the vth dimension, and the jth column (ordinal/predictor) category along the sth dimension of a correspondence plot, their principal coordinate is defined by

$$f_{iv} = \sum_{s=1}^{S} a_{is} \lambda_{s(v)}$$

$$g_{js} = \sum_{v=1}^{J-1} \beta_{jv} \lambda_{s(v)} \, ,$$

respectively, for $s = 1, \ldots, S$ and $v = 1, \ldots, J-1$. The properties of these coordinates are very similar to those described in Section 4.5.2 for doubly ordered correspondence analysis. That is, these coordinates are centred at the origin of the plot, and along each dimension, and the total variation can be expressed in terms of these coordinates by

$$\tau_{\text{num}} = \sum_{i=1}^{I} \sum_{v=1}^{J-1} f_{iv}^2 = \sum_{j=1}^{J} \sum_{s=1}^{S} p_{\bullet j} g_{js}^2 \, .$$

Another property that these coordinates share with those derived for simple ordinal correspondence analysis is that each dimension has two explained inertia values; one for the row variable and one for the column variable. That is, the value of the explained inertia for the row categories along the vth dimension is

$$\lambda_{\bullet(v)}^2 = \sum_{i=1}^{I} f_{iv}^2$$

while the explained inertia of the column categories along the sth dimension is

$$\lambda_{s(\bullet)}^2 = \sum_{j=1}^{J} p_{\bullet j} g_{js}^2 \, .$$

Note that even in the case where $s = v$, these two inertia values may well be different since they are defined in terms of the $\lambda_{s(v)}^2$ values. For this reason, we shall describe the benefits of constructing a *biplot* for the visualisation of the asymmetric association between the row and column variables.

5.5 The Biplot

5.5.1 An Overview

Recall that in the previous chapters we described that one can construct a *row isometric biplot* or a *column isometric biplot*, depending on which variable an analyst wishes to focus on. We can also construct these biplots when a singly ordered non-symmetrical correspondence analysis is performed on a two-way contingency table. Here we describe how these biplots can be constructed, interpreted, and applied to the shoplifting data summarised in Table 1.3.

5.5.2 Column Isometric Biplot

To visually summarise the asymmetric association between the *Age* and *Item* variables using non-symmetrical correspondence analysis, we construct a column isometric biplot. For such a biplot, the *i*th row and *j*th column categories along the *v*th dimension of a correspondence plot is defined by

$$\tilde{f}_{is} = a_{is} \tag{5.7}$$

$$\tilde{g}_{js} = \sum_{v=1}^{J-1} \beta_{jv} \lambda_{s(v)} , \tag{5.8}$$

respectively, for $s = 1, \ldots , S$. For the column isometric biplot, the row categories are depicted as a projection from the origin to their standard coordinates defined by the elements of the left singular vectors, while the column categories are portrayed as points at their principal coordinates. By using these coordinates, the $\pi_{i|j}$ values defined by Eq. (5.1) can be fully reconstituted by

$$\pi_{i|j} = \sum_{s=1}^{S} \tilde{f}_{is} \tilde{g}_{js} \tag{5.9}$$

and is the inner product of the coordinates defined by Eq. (5.7) and Eq. (5.8).

Interestingly, the column principal coordinates can be defined in terms of $\pi_{i|j}$ such that

$$\tilde{g}_{js} = \sum_{i=1}^{I} \pi_{i|j} a_{is} .$$

Therefore, the column isometric biplot coordinates for the row and column categories can be expressed in terms of the elements of the left singular vectors. This means that the configuration of points in a such a visual display is identical to the configuration of points in a column isometric biplot obtained from the (nominal) non-symmetrical correspondence analysis of a two-way contingency table.

5.5.3 Column Isometric Biplot of the Shoplifting Data

Figure 5.2 is the column isometric biplot obtained from performing a singly ordered non-symmetrical correspondence analysis on the shoplifting data summarised in

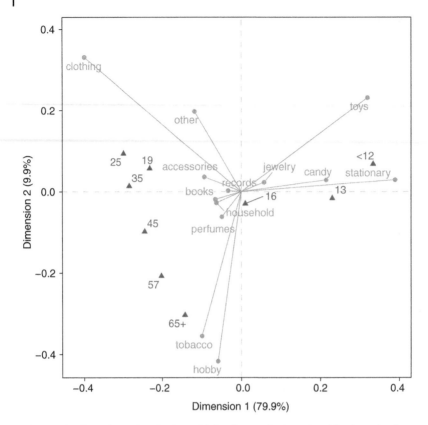

Figure 5.2 Two-dimensional column biplot from a singly ordered (and nominal) non-symmetrical correspondence analysis of Table 1.3 with beta scaling ($\beta = 1.5$).

Table 1.3; to provide a clear interpretation of the row/column interactions that are present, beta scaling has been applied where $\beta = 1.5$. For the construction of this biplot, the coordinates used to visually summarise the asymmetric association between the *Item* and *Age* variables are those defined by Eq. (5.7) and Eq. (5.8), respectively. Note that this figure is also the same column isometric biplot that one obtains by applying to Table 1.3 the nominal version of non-symmetrical correspondence analysis described in Chapter 3.

Figure 5.2 shows that there are clear differences, and similarities, in the way that each *Age* category and *Item* category contribute to the asymmetric association structure. On the right we see the three youngest age groups and the items they are most likely to steal while the right-hand side of Figure 5.2 comprises of the remaining age groups. For example, for the configuration on the right of the correspondence plot, those male perpetrators who are aged 14 years or less (< 12 and 13) are most likely to steal *candy* and *stationary*. The item *toys* also appears on this side of the plot and shows that they are more likely to be stolen by males who are aged less than 12 years than any other age group. We can also see that the *Age* group 16 lies closest to the origin of the correspondence plot. Therefore, this particular category plays a weak role in defining the asymmetric association, at least compared to the other *Age* groups. These findings are consistent with our discussion of the the age group profiles depicted in Figure 5.1.

For the configuration of points on the left-hand side of Figure 5.2, it shows that those who are aged 65+ have a higher propensity to steal *tobacco* and *hobby* items than any other item. It also shows that *clothing* appears to be an important contributor to the asymmetric association and is more likely to be stolen by 21 to 29 year olds (25) than any other age group. There are many items, including *perfumes, household, books, accessories, records* and *jewelery* that are closer to the origin than any other stolen item which suggests their role in defining the asymmetric association between the two variables of Table 1.3 is not as dominant as other *Item* categories.

Suppose we now focus our attention on the quality of the visual summary that Figure 5.2 gives of the asymmetric association between the *Age* and *Item* variables. Since Figure 5.2 is identical to the correspondence plot obtained from applying (nominal) non-symmetrical correspondence analysis to Table 1.3, the quality of this two-dimensional correspondence plot is assessed in terms of the singular values of the matrix of $\pi_{i|j}$ values. Therefore, from such values, the first dimension of Figure 5.2 describes 79.9% of the association while the second dimension accounts for 10.0%. So, Figure 5.2 provides a very good visual depiction of the association with 89.9% of the total inertia (quantified by τ_{num}) explained; note that these percentages are also summarised in the last column of Table 5.4.

5.5.4 Row Isometric Biplot

Suppose we now visually summarise the asymmetric association between the row and column variables of a two-way contingency table by depicting the row categories as points defined by their principal coordinate and the column categories as projections from the origin to their standard coordinates (defined using the elements of the column orthogonal polynomials). Doing so produces a *row isometric biplot*. For such a biplot, the ith row and jth column categories along the vth dimension is defined by

$$\tilde{f}_{iv} = \sum_{s=1}^{S} a_{is} \lambda_{s(v)}$$

$$= \sum_{j=1}^{J} p_{\bullet j} \pi_{i|j} \beta_{jv} \tag{5.10}$$

$$\tilde{g}_{jv} = \beta_{jv} , \tag{5.11}$$

respectively, for $v = 1, \ldots, J - 1$. From the definition of these coordinates, and Eq. (5.1), the $\pi_{i|j}$ values can be fully reconstituted by

$$\pi_{i|j} = \sum_{v=1}^{J-1} \tilde{f}_{iv} \tilde{g}_{jv} \tag{5.12}$$

and is the inner product of the coordinates defined by Eq. (5.10) and Eq. (5.11).

5.5.5 Row Isometric Biplot of the Shoplifting Data

Figure 5.3 is the two-dimensional row isometric biplot obtained from performing a singly ordered non-symmetrical correspondence analysis on the data summarised in Table 1.3.

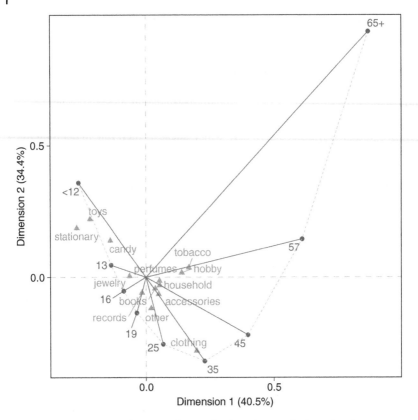

Figure 5.3 Two-dimensional row isometric biplot from a singly-ordered non-symmetrical correspondence analysis of Table 1.3 with beta scaling ($\beta = 4$).

Therefore, the coordinates used to visually summarise the asymmetric association between the *Item* and *Age* variables are those defined by Eq. (5.10) and Eq. (5.11), respectively. Since the row and column coordinates used to construct this biplot are expressed in terms of the orthogonal polynomials, its configuration of points will not be the same as the configuration obtained if a row isometric biplot is constructed from a (nominal) non-symmetrical correspondence analysis. As we shall describe, nor is the interpretation of the dimensions the same as those obtained from performing a (nominal) non-symmetrical correspondence analysis. However, the general interpretation of the association is still very apparent and can be described in terms of differences in the mean and variation of the profiles.

Unlike the column isometric biplot of Figure 5.2, Figure 5.3 shows the differences in the mean values of the *Age* profiles and in their variation. The first dimension of Figure 5.3 reflects the differences in the mean values of the *Age* profiles when taking into consideration the ordered structure of the *Age* categories. Similarly, its second dimension describes the differences in the variation of these profiles.

When the first two dimensions are used to construct Figure 5.3, the resulting configuration of the *Age* categories has an obvious parabolic shape. This is because the column orthogonal polynomials are used to define the column standard coordinates. Therefore, their depiction along the first dimension of biplot is in the same order as the sequence of

Age categories. It shows that the average age of the perpetrators is somewhere between 19 and 25 years (in fact, the mean age of the perpetrators is 21.2 years) and that the profiles of the oldest perpetrators (aged 65+ years) have the greatest variability. However, the standard coordinates of the *Age* categories do not reflect the association this variable has with the *Item* variable. Although, conclusions of this association can be made by observing the distance of the point of each *Age* group from the projection that the *Item* categories have from the origin. For example, Figure 5.3 shows that male perpetrators that are aged less than 12 years (< 12), or who are between 12 and 14 years (13) are more likely to steal *toys*, *candy* and *stationary* than any other item; note that no other age group has a high propensity to steal these items. However, the profile of those perpetrators aged less than 12 years of age (< 12) shows a larger variation than those between 12 and 14 years (13). This can be observed by noting that the *Age* category < 12 has a larger second coordinate than the category 13.

Figure 5.3 also shows that, given that someone who is aged 50 to 64 years of age (57), they are more likely to steal *tobacco* and *hobby* items than any other item. Similarly, for those in the 19 and 25 *Age* categories, they have a strong propensity to steal *other* items and *records*. Furthermore, Figure 5.3 shows that *clothing*, which has a dominant impact on the asymmetric association, appears more likely to be stolen by perpetrators aged 30 to 39 years (35) than by those in the 25 *Age* category. The variation of these age groups is not so large when compared to the variation of the oldest age group.

Note that these conclusions are exactly the same as those found by observing the configuration of points in Figure 5.2, which is as it should be. However, the dimensions of Figure 5.3 provide a clear interpretation of the reasons for why the *Item* categories are similar, or different. Such an interpretation is certainly clearer than the dimensions of Figure 5.2 provide for the profiles of the *Age* categories. To demonstrate this, we now turn our attention to the quality of the row isometric biplot of Figure 5.3. In doing so, we provide an interpretation of the explained inertia value for its two dimensions.

In general, the value of the explained inertia along the vth dimension of a row isometric biplot of the type given by Figure 5.3 is quantified by

$$\sum_{j=1}^{J} p_{\bullet j} g_{jv}^2 = \sum_{v=1}^{J-1} \lambda_{s(v)}^2 \, .$$

For example, from the summary of the $\lambda_{s(v)}$ values in Table 5.4, the explained inertia along the first dimension ($v = 1$) is

$$\sum_{s=1}^{S=8} \lambda_{1(v)}^2 = \lambda_{1(1)}^2 + \lambda_{2(1)}^2 \cdots + \lambda_{8(1)}^2$$

$$= (-0.115)^2 + (-0.041)^2 + \ldots + (0.000)^2$$

$$= 0.015 \, .$$

Since $\tau_{\text{num}} = 0.0375$, this dimension therefore reflects

$$100 \times \frac{0.015}{0.0375} = 40.4\%$$

of the predictability between the two variables; note that this is exactly the percentage value we get in the last row of the first column of Table 5.4. Similarly, the explained inertia value along the second dimension ($v = 2$) of Figure 5.3 is

$$\sum_{s=1}^{S=8} \lambda_{s(2)}^2 = \lambda_{1(2)}^2 + \lambda_{2(2)}^2 \cdots + \lambda_{8(2)}^2$$
$$= (0.108)^2 + (-0.026)^2 + \ldots + (0.000)^2$$
$$= 0.013$$

and reflects

$$100 \times \frac{0.013}{0.0375} = 34.6\%$$

of the predictability of the *Item* categories given the *Age* of the perpetrators. Therefore, Figure 5.3 visually summarises 75% of the association between the *Age* and *Item* variables of Table 1.3. These values demonstrate that, while providing a visual summary of the asymmetric association, the differences in the profiles of the *Item* categories are best described by the differences in their mean values, followed closely by the differences in their variation. So Figure 5.3 not only provides a good visual summary of the asymmetric association between the *Age Item* variable, but it does so by being explicit as to the reasons why there are differences in the profiles.

5.5.6 Distance Measures and the Row Isometric Biplots

While the distance of a standard coordinate from the projection of a principal coordinate (from the origin) is an important feature of isometric biplots, so too is the distance of principal coordinates from each other and the origin. To discuss this feature, we confine our attention to examining such distances for the row isometric biplot.

Suppose that we apply singly ordered non-symmetrical correspondence analysis to a two-way contingency table with ordered predictor column categories and nominal response row categories; like we saw for our analysis of Table 1.3. Then, the squared Euclidean distance of the ith principal coordinate, $\tilde{f}_{iv}$, from the origin of an optimal row isometric biplot is

$$d_I^2(i, \ 0) = \sum_{v=1}^{J-1} (\tilde{f}_{iv} - 0)^2$$
$$= \sum_{v=1}^{J-1} \tilde{f}_{iv}^2 \ .$$

Therefore, the total inertia from this analysis can be expressed in terms of this distance by

$$\tau_{\text{num}} = \sum_{i=1}^{I} d_I^2(i, \ 0) \ .$$

This result shows that row (response) points located as a distance from the origin play a more dominant role in defining the asymmetric association structure between the variables than row (response) points that are close to the origin. This means that row points located close to the origin are deemed to be poorly "predicted" compared to those row points located further away from the origin.

Suppose we now consider the squared-Euclidean distance between the ith and i'th row principal coordinates so that

$$d_I^2(i,\ i') = \sum_{v=1}^{J-1} (\tilde{f}_{iv} - \tilde{f}_{i'v})^2 \ .$$

Then this distance is also just the difference between the centred profiles of these two row categories since

$$d_I^2(i,\ i') = \sum_{j=1}^{J} p_{\bullet j} \left(\frac{p_{ij}}{p_{\bullet j}} - \frac{p_{i'j}}{p_{\bullet j}} \right)^2$$

$$= \sum_{j=1}^{J} p_{\bullet j} \left[\left(\frac{p_{ij}}{p_{\bullet j}} - p_{i\bullet} \right) - \left(\frac{p_{i'j}}{p_{\bullet j}} - p_{i'\bullet} \right) \right]^2 \ .$$

Therefore, row (response) categories that have similar profiles will share similar positions in the optimal correspondence plot. This also means that row (response) categories that have very different profiles will be located at some distance from each other. Therefore, by using the row principal, or biplot, coordinates defined by Eq. (5.10), singly ordered non-symmetrical correspondence preserves the *property of distribution equivalence* described by Lebart et al. (1984, p. 35), Greenacre (1984, p. 95) and Beh and Lombardo (2014, Section 4.6). This property is also preserved for the *doubly ordered non-symmetrical correspondence analysis* technique described by Lombardo, Beh and D'Ambra (2007) and Beh and Lombardo (2014, Section 7.3); such a variant is for a two-way contingency table where the variables are both ordered and have an asymmetric association structure.

5.6 Some Final Words

Recall that, by applying a non-symmetrical correspondence analysis to a singly ordered two-way contingency table, we can reconstitute the $\pi_{i|j}$ values based on the information depicted in a two-dimensional isometric biplot; see Eq. (5.9) and Eq. (5.12). Similar results can also be obtained for the analysis of a table where both the row and column variables consist of ordered categories. Even though we have focused our attention solely on the first dimensions of the isometric biplot, it is sometimes helpful to know that dimensions beyond the second can be used to detect profile trends or changes in the predictability of the *Age* categories in terms of third- and higher-order moments. For singly ordered symmetrical and non-symmetrical analysis we do need to determine whether the row or column variable consists of ordered categories since this helps determine the type of biplot that will be constructed. The distinction is made in terms of how the dimensions are to be interpreted and whether it is the differences in the row profiles or the column profiles that will be assessed in terms of their mean value or variation.

For the analysis of Table 1.3, the first two components (reflecting differences in the mean and variation of a profile) are sufficient to assess the predictability of the *Item* that is stolen given the *Age* of the perpetrator. However, improvements in assessing this predictability can be made by including a third dimension to our visual summary. In fact, the inclusion of a higher dimension provides a much better visual assessment of the prediction of *clothing* and *stationary* than does the inclusion of the third dimension. This is because, for those row categories the differences in their profiles are better explained in terms of their kurtosis than their skewness. However, we shall not discuss this aspect of the analysis further here since it does add another layer of detail to how one may interpret the profile differences for the row and column categories as well as the asymmetric association structure between their variables. Therefore, we invite the interested reader, at their own leisure, to ponder on this aspect of correspondence analysis further.

Part III

Analysis of Multiple Categorical Variables

6

Multiple Correspondence Analysis

6.1   Introduction

Earlier chapters of this book have discussed a variety of correspondence analysis techniques for two categorical variables. In particular, we provided an overview of how correspondence analysis can be applied to analyse nominal and ordinal categorical variables that are either symmetrically or asymmetrically associated. We now describe some common approaches that can be used for simultaneously visualising the association between multiple categorical variables by focusing on the analysis of only three variables. The data structures that are typically studied are contingency tables formed from the cross-classification of three or more variables and the correspondence analysis of these tables is referred to as *multiple correspondence analysis*, often initialised to MCA. Much of the literature on multiple correspondence analysis is confined to the analysis of multiple symmetrically associated nominal variables and so we shall confine our attention to this case. The availability of tools to analyse a mix of nominal and ordinal variables that are symmetrically or asymmetrically associated using the techniques we describe in this chapter is very limited. Although in Chapter 7 we shall discuss such methods by using an approach to correspondence analysis called *multi-way correspondence analysis*, which can be initialised to MWCA.

The underlying principles of multiple correspondence analysis are akin to those of principal component analysis, but focus on categorical data rather than numerical data. It is also commonly viewed as a form of metric multi-dimensional scaling; see, for example, de Leeuw (1984), Hoffman and de Leeuw (1992) and Hoffman et al. (1994). Thus, the groundwork of multiple correspondence analysis has been invented, and reinvented, many times. In fact, multiple correspondence analysis is known by many names, including *optimal scaling*, *optimal scoring*, *dual scaling* and *homogeneity analysis*. In our discussion here we shall simply refer to it as *multiple correspondence analysis*.

The heart of multiple correspondence analysis lies in recoding a multi-way contingency table so that it is in the form of a two-way table. This can be done in a variety of ways and, despite the popularity of the approaches we shall describe, there is always some loss of information in the association between the variables in doing so. In this chapter we confine the application of multiple correspondence analysis to the visual summary of the association between three categorical variables, although it can be applied to a much larger sized table without loss of generality. The most common ways of transforming a three-way (or more generally, a multi-way) contingency table into a two-way table involve

- Recoding the contingency table into a matrix of 0's and 1's called an *indicator matrix*. In constructing such a matrix, *crisp coding* is used for each individual that is sampled so that a coding of "1" reflects that an individual exhibits a category response while a "0" means that they do not. Sometimes such "crispness" in coding is not straight forward. Sometimes *fuzzy coding* is used (van Rijckevorsel 1987) so that scores (or *probability values*) are assigned to each category where the scores (or probabilities) across the categories of a variable sum to 1.
- Recoding the contingency table into a super-diagonal matrix called a *Burt matrix*, in honour of Sir Cyril Burt (Burt 1950).
- Or, stacking two, or more, of the variables that are cross-classified into a multi-way contingency table; see, for example, Weller and Romney (1990).

We now turn our attention to briefly describing these three coding methods and how they can be used to perform multiple correspondence analysis.

6.2 Crisp Coding and the Indicator Matrix

6.2.1 Crisp Coding

Crisp coding refers to the coding of an individual in a sample such that, for each category of a variable, either a 0 or a 1 is recorded; a 1 denotes that the individual is observed to have the characteristic described by a category, otherwise it is zero. For each variable, the coding is completely disjunctive such that only one 1 is recorded. For example, for three variables, an individual will have a 1 coded three times (one for each variable) and many zeros. For a sample of individuals, this coding forms an indicator matrix.

6.2.2 The Indicator Matrix

We now adopt a more formal description of the indicator matrix. Suppose we have M categorical variables, $X_1, \ldots, X_l, \ldots, X_M$, that are cross-classified to form a M-way contingency table. Denote the number of categories for variable X_l by k_l, where $l = 1, 2, \ldots, M$. Also, let the number of observations that are cross-classified be denoted by n. We can then summarise the crisp coding of variable X_l using a $n \times k_l$ indicator matrix $\mathbf{Z}_l$, where each row of $\mathbf{Z}_l$ sums to one and its column totals are just the marginal frequencies of that variable. We then concatenate all of the M indicator matrices thereby forming a super-indicator matrix denoted by

$$\mathbf{Z} = [\mathbf{Z}_1 \quad \mathbf{Z}_2 \quad \ldots \quad \mathbf{Z}_l \quad \mathbf{Z}_M] \; .$$

For completeness, we define

$$J = \sum_{l=1}^{M} k_l$$

to be the total number of categories across the M variables of the contingency table.

There are many advantages to using the crisp coding; the interested reader is directed to, for example, van Rijckevorsel (1987) who provides an overview of many of them. Two such advantages include its simplicity which lends itself nicely to computations and the geometric nature of the resulting correspondence plot.

6.2.3   Crisp Coding and the Alligator Data

To demonstrate the application of crisp coding, consider again the $2 \times 5 \times 4$ three-way contingency table of Table 1.4 which summarises the *Size*, *Lake* and primary *Food* source of 219 alligators in Florida as described by Agresti (2002). Since there are three variables with a total of $J = 2 + 5 + 4 = 11$ categories, crisp coding will yield an indicator matrix of size 219×11; a partial table obtained from the crisp coding of Table 1.4 is given by Table 6.1. For each row (reflecting the classifications made of each of the alligators that were studied), a 1 is entered into the category where the alligator was classified and a 0 where a classification was not made. For example, the first row of Table 6.1 represents a *Small* alligator in Lake *Hancock* whose primary food choice was found to be *Fish*. As a result, we see that the first row has a 1 for the *Size* category *Small*, a 1 for the *Food* category *Fish* and a 1 for the *Lake* category *Hancock*. All other values in this row are zero. Therefore, a multiple correspondence analysis of the data summarised in Table 1.4 can be made by performing a simple correspondence analysis on Table 6.1. The data array given by Table 6.1 is commonly referred to as the *super-indicator matrix* of Table 1.4. Refer to Chapter 2 for an introductory description of simple correspondence analysis.

In general, when performing multiple correspondence analysis via the $n \times J$ indicator matrix **Z**, the total inertia can be easily calculated by

$$\frac{X_Z^2}{n} = \frac{J}{M} - 1 . \tag{6.1}$$

See, for example, Greenacre (1984, p. 139) and Weller and Romney (1990, p. 67). Therefore, the total inertia of the indicator matrix falls within the interval $0 \le X_Z^2/n \le \min(n, J) - 1$.

Table 6.1   Indicator table from the crisp coding of Table 1.4.

	Size		Primary Food Choice					Lake			
Alligator	Small	Large	Fish	Invertebrate	Reptile	Bird	Other	Hancock	Oklawaha	Trafford	George
1	1	0	1	0	0	0	0	1	0	0	0
2	1	0	1	0	0	0	0	1	0	0	0
⋮	⋮	⋮	⋮	⋮	⋮	⋮	⋮	⋮	⋮	⋮	⋮
23	1	0	1	0	0	0	0	1	0	0	0
24	1	0	1	0	0	0	0	0	1	0	0
⋮	⋮	⋮	⋮	⋮	⋮	⋮	⋮	⋮	⋮	⋮	⋮
28	1	0	1	0	0	0	0	0	1	0	0
29	1	0	1	0	0	0	0	0	0	1	0
⋮	⋮	⋮	⋮	⋮	⋮	⋮	⋮	⋮	⋮	⋮	⋮
300	0	1	0	0	0	0	1	0	0	0	1

For example, for Table 1.4, since $M = 3$ (for the three variables being analysed) and $J = 11$ (the number of categories amongst the three variables) then the total inertia of the indicator matrix of Table 6.1 is

$$\frac{X_Z^2}{n} = \frac{2 + 5 + 4}{3} - 1 = 2.666$$

and falls within the range $0 \leq X_Z^2/n \leq \min(219, 11) - 1 = 10$. Therefore, one may imply from this result that, since the total inertia appears to be fairly small in comparison to what it could be, we may conclude that the association between the three variables is fairly weak. However, this is not the case. In fact, as we showed in Section 1.6.5, the Pearson chi-squared statistic of Table 1.4 is 85.492 and, with 31 degrees of freedom, has a p-value that is less than 0.0001. Therefore, there is a statistically significant association between the variables of this three-way contingency table.

6.2.4 Application of Multiple Correspondence Analysis using the Indicator Matrix

Now that we have applied crisp-coding to Table 1.4 and constructed the indicator matrix that is (partly) summarised in Table 6.1 we can visually study the association between the three variables by performing multiple correspondence analysis. This can be done for Table 1.4 by applying a simple correspondence analysis to the 219×11 table of Table 6.1. Note that, when performing such an analysis, the optimal correspondence plot will consist of $S = \min(219, 11) - 1 = 10$ dimensions making it a very large and complex space to work with. However, Greenacre (1984, p. 139) points out that the optimal number of dimensions from performing multiple correspondence analysis using the indicator matrix will actually consist of $J - M = 11 - 3 = 8$ dimensions. So, reducing this 8-dimensional optimal space down to a low-dimensional space that consists of two or (at most) three dimensions will generally mean that only a small portion of the association (quantified using the total inertia) will be visually displayed in the resulting correspondence plot. For the alligator data of Table 1.4, Figure 6.1 provides a two-dimensional correspondence plot from the multiple correspondence analysis of its indicator matrix; refer to Section 2.5.2 for an introductory description of how to construct a correspondence plot from a simple correspondence analysis of a two-way contingency table.

Figure 6.1 shows that the *Large* alligators in Lake *Trafford* (Tr) eat a mix of *Fish*, *Reptiles* and *Invertebrates* as their primary source of, while the *Small* alligators in Lake *George* (Ge) feed off *Fish* and *Invertebrates*. One may note that the position of *Fish* is in close proximity to the origin of Figure 6.1 showing that, relative to other points, this category does not appear to play a dominant role in defining the association between the three variables of Table 1.4. Figure 6.1 also shows that for the other lakes in the study (*Handcock* (Ha) and *Oklawaha* (Ok)) there is an even mix of *Small* and *Large* alligators. In particular, the alligators in Lake *Hancock* (Ha) eat primarily *Birds* and *Other* animals, while those in Lake *Oklawaha* (Ok) have a stronger taste for *Reptiles* and *Invertebrates* than any of food type but do not appear to have a liking for *Birds*. Furthermore, ignoring for the moment the alligator's lake of residence and focusing solely on the association between *Size* and *Food*, we observe that the *Large* alligators eat mainly *Reptiles* and *Birds*, while the *Small* alligators feed primarily on *Invertebrates* and *Fish*. Such results suggest that there is indeed some variety of size and

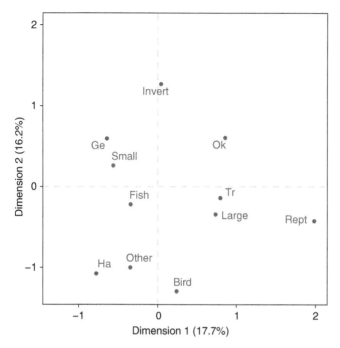

Figure 6.1 Two-dimensional correspondence plot from the multiple correspondence analysis, using the indicator matrix, of Table 1.4.

food source amongst the alligators in the Floridian lakes and Figure 6.1 provides a visual glimpse into how these three variables are associated.

However, like we discussed in the previous chapters, one important issue that needs to be addressed is the quality of the correspondence plot given by Figure 6.1. Its quality can be assessed by determining the singular value, λ_s^Z for the sth dimension, for $s = 1, \ldots, 8$. For Table 1.4 the square of these values is

$$(\lambda_1^Z)^2 = 0.471, \; (\lambda_2^Z)^2 = 0.432, \; (\lambda_3^Z)^2 = 0.390, \; (\lambda_4^Z)^2 = 0.346,$$

$$(\lambda_5^Z)^2 = 0.316, \; (\lambda_6^Z)^2 = 0.286, \; (\lambda_7^Z)^2 = 0.254, \; (\lambda_8^Z)^2 = 0.171$$

and are the explained inertia values along each of the dimensions of the 8-dimensional optimal correspondence plot. Therefore, the total inertia of the indicator matrix of Table 1.4 is

$$\frac{X_Z^2}{n} = \sum_{s=1}^{8} (\lambda_s^Z)^2$$
$$= 0.471 + 0.432 + \ldots + 0.171$$
$$= 2.666$$

and is identical to the total inertia we saw using Equation (6.1). These explained inertia values are summarised in Table 6.2 as is their percentage contribution to the total inertia and their cumulative percentage contribution of each dimension. We can see that the first dimension of Figure 6.1 accounts for 17.7% of the total inertia while a further 16.2% is accounted for by the second dimension. Therefore, Figure 6.1 visually describes only 33.9% of the total inertia – a very low quality figure indeed. So one may consider this

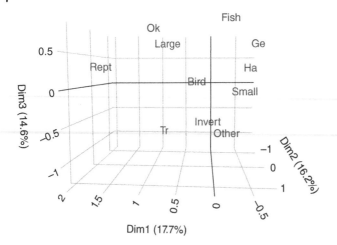

Figure 6.2 Three-dimensional correspondence plot from the multiple correspondence analysis, using the indicator matrix, of Table 1.4.

two-dimensional correspondence plot to be a very poor visual summary of the association between the *Lake*, *Size* and *Food* variables of Table 1.4.

Things do not improve greatly even if we were to add a third dimension to this display; see Figure 6.2 for a three-dimensional correspondence plot of the association. The third variable accounts for 14.6% of the total inertia so that Figure 6.2 visually summarises about 48.5% of the association. Despite this, the presence of the third dimension reveals, for example, that the interaction between Lake *Trafford* and *Large* sized alligators is not as strong as Figure 6.1 suggests. Nor is the interaction between Lake *Hancock* and *Other* sources of food as close as what the two-dimensional plot might suggest. However, *Small* sized alligators still appear to be closely associated with Lake *George* with the third dimension included. With the inclusion of the third dimension, the association structure is quite different to what Figure 6.1 suggests. With the information reflected in higher dimensions not being displayed in either the two- or three-dimensional correspondence plots, there is no doubt that the true nature of the association has not been revealed with the construction of Figure 6.1 and Figure 6.2.

The issue of the quality of the two-(or three-)dimensional correspondence plot is a very important one since it provides a visual summary of the association between variables that would typically require many more dimensions to obtain an acceptable display. In many practical applications of multiple correspondence analysis, the quality issue is one that is often ignored, either unintentionally (because an analyst is not aware of its importance) or because analysts do not wish to reveal how poor their display is. Irrespective of the cause of such an omission, ignoring the quality is much the same as ignoring the R^2 value, for example, when performing a linear regression analysis (something most analysts would consider unthinkable!). In practice, using the indicator matrix to perform multiple correspondence analysis will generally produce poor quality two- and three-dimensional displays because of the relatively low proportions that the explained inertia values contribute to the total inertia; we can see, for example, that the highest percentage (being for the first dimension of course) from Table 6.2 is only 17.7% and so all other dimensions will have even smaller percentage contribution values. This is especially so even if the number of variables being studied is small and the number of categories within each variable is also

Table 6.2 Inertia values from the correspondence analysis of the indicator matrix of Table 1.4.

Dimension	Explained Inertia	Percentage Inertia	Cumulative Inertia
1	0.471	17.669	17.669
2	0.432	16.196	33.865
3	0.390	14.612	48.477
4	0.346	12.979	61.455
5	0.316	11.837	73.293
6	0.286	10.744	84.037
7	0.254	9.536	93.572
8	0.171	6.428	100.000
Total	2.666	100.000	

small. Greenacre (1990), for example raised this point as well but pointed out that these low inertia values can be "corrected" by using an adjustment proposed by Benzécri (1979). Such an adjustment to the sth singular value can be made such that

$$\text{adj}(\lambda_s^Z) = \left[\frac{M}{M-1} \left(\lambda_s^Z - \frac{1}{M} \right) \right]^2 .$$

However Greenacre (1990, pg 250) notes that such an adjustment came with "a rather unconvincing justification".

On a final note concerning crisp coding, adopting a 0–1 coding scheme means that an individual (or unit) is assigned exclusively to only one category for each variable. When there is uncertainty as to which category an individual/unit belongs, one can consider instead assigning scores (or probability values) that reflect a particular "fuzzy rule" for group membership. That is, rather than assigning an individual/unit a 0 or 1 for each category, an alternative strategy is to assign multiple score values (that sum to 1) for each variable. Such a coding procedure is referred to as *fuzzy coding* and can be undertaken in a number of ways. Due to the flexibility in specifying how an individual/unit should be classified, a fuzzy coding scheme is generally preferred to crisp coding, which often gives rise to discontinuity issues. Fuzzy coding ensures that observations close to the boundary that separates two adjacent categories are not very different in their recoded values. It means that distances between adjacent categories are reduced and the interaction between the categories is made clearer. Van Rijckevorsel (1987) discusses that the first analysts to introduce the basic idea of fuzzy coding were Bordet (1973), Ghermani et al. (1977) and Guitonneau and Roux (1977). Since then, fuzzy coding has been introduced throughout the French literature by linking its use with multiple correspondence analysis (Gallego 1980; Le Foll 1979), the development of theoretical probabilistic properties of the chi-squared statistic (Martin 1980) and methods for dealing with discrimination problems (Gautier and Saporta 1982). Similarly, de Leeuw et al. (1981), van Rijckevorsel (1987) and Gifi (1990) discussed the role of fuzzy coding in a correspondence analysis context by confining their attention to B-spline functions. Greenacre (1984) also discusses different types of fuzzy coding strategies that can be considered as alternatives to crisp coding.

6.3 The Burt Matrix

To help improve the quality of a two-(or three-)dimensional correspondence plot obtained from performing multiple correspondence analysis using the indicator matrix, consider the following approach. Suppose, for the sake of simplicity, that the contingency table, $\mathbf{N}$, is formed from the cross-classification of two categorical variables, X_1 and X_2. When using crisp coding, we obtain an indicator matrix for variable X_1 which we denote by $\mathbf{Z}_1$. Similarly, for variable X_2, its indicator matrix is denoted as $\mathbf{Z}_2$. A feature of these indicator matrices is that their product gives the original two-way contingency table, $\mathbf{N}$ such that

$$\mathbf{N} = \mathbf{Z}_1^T \mathbf{Z}_2 .$$

Suppose we now consider three categorical variables X_1, X_2 and X_3 so that cross-classifying them gives a three-way contingency table. We define the indicator matrix of each of these variables by $\mathbf{Z}_1$, $\mathbf{Z}_2$ and $\mathbf{Z}_3$, respectively. Therefore, the concatenation of these indicator matrices yields

$$\mathbf{Z} = [\mathbf{Z}_1 \quad \mathbf{Z}_2 \quad \mathbf{Z}_3]$$

so that we obtain the *Burt matrix*

$$\begin{aligned}
\mathbf{B} &= \mathbf{Z}^T \mathbf{Z} \\
&= \begin{pmatrix} \mathbf{Z}_1^T \mathbf{Z}_1 & \mathbf{Z}_1^T \mathbf{Z}_2 & \mathbf{Z}_1^T \mathbf{Z}_3 \\ \mathbf{Z}_2^T \mathbf{Z}_1 & \mathbf{Z}_2^T \mathbf{Z}_2 & \mathbf{Z}_2^T \mathbf{Z}_3 \\ \mathbf{Z}_3^T \mathbf{Z}_1 & \mathbf{Z}_3^T \mathbf{Z}_2 & \mathbf{Z}_3^T \mathbf{Z}_3 \end{pmatrix} \\
&= \begin{pmatrix} \mathbf{D}_1 & \mathbf{N}_{12} & \mathbf{N}_{13} \\ \mathbf{N}_{12}^T & \mathbf{D}_2 & \mathbf{N}_{23} \\ \mathbf{N}_{13}^T & \mathbf{N}_{23}^T & \mathbf{D}_3 \end{pmatrix} .
\end{aligned} \tag{6.2}$$

Here, $\mathbf{D}_m$ is the diagonal matrix of marginal cell frequencies of the mth variable, for $m = 1, 2, 3$. Similarly, $\mathbf{N}_{ab}$ is the two-way contingency table formed by cross-classifying variables X_a and X_b, for $a \neq b$, and is formed by aggregating across the remaining variable. For M categorical variables, where $M \gg 3$, the Burt matrix, $\mathbf{B}$, can be obtained by

$$\mathbf{B} = \mathbf{Z}^T \mathbf{Z}$$

where

$$\mathbf{Z} = [\mathbf{Z}_1 \quad \mathbf{Z}_2 \quad \cdots \quad \mathbf{Z}_M] .$$

In cases where the sample size is very large, the indicator matrix can consist of hundreds, thousands or even many more rows. This is one practical reason why performing multiple correspondence analysis via the indicator matrix is not advisable – one can run into serious computational issues if the sample size is VERY large (Iodice D'Enza, Groenen and van de Velden, 2020). Instead, one may perform multiple correspondence analysis by applying simple correspondence analysis to the Burt matrix instead of the indicator matrix. For example, Table 6.3 is the Burt matrix for Table 1.4, derived using its indicator matrix given (in part) by Table 6.1. This table is a "super" two-way contingency table consisting of nine identifiable tables for each pair of variables, the form of which may be compared with Equation (6.2).

Table 6.3 Burt table of Table 1.4.

Alligator	Size		Primary Food Choice					Lake			
	Small	Large	Fish	Invertebrate	Reptile	Bird	Other	Hancock	Oklawaha	Trafford	George
Small	124	0	49	45	6	5	19	39	20	24	41
Large	0	95	45	16	13	8	13	16	28	29	22
Fish	49	45	94	0	0	0	0	30	18	13	33
Invertebrate	45	16	0	61	0	0	0	4	19	18	20
Reptile	6	13	0	0	19	0	0	3	7	8	1
Bird	5	8	0	0	0	13	0	5	1	4	3
Other	19	13	0	0	0	0	32	13	3	10	6
Hancock	39	16	30	4	3	5	13	55	0	0	0
Oklawaha	20	28	18	19	7	1	3	0	48	0	0
Trafford	24	29	13	18	8	4	10	0	0	53	0
George	41	22	33	20	1	3	6	0	0	0	63

The top-left table is just the 2×2 table whose diagonal elements consist of the marginal frequencies of the *Size* variable. Similarly, the diagonal bottom-right 4×4 table consists of the marginal frequencies for the *Lake* variable. Therefore, the sum of the marginal frequencies summarised in the three diagonal tables of Table 6.3 gives a total of $n = 219$.

The top-right table of Table 6.3 is the 2×4 contingency table obtained by cross-classifying only the *Size* and *Lake* variables, while the bottom-left table is the transpose of this matrix. The two-way tables formed from the pairwise combinations of the three variables, and their transpose, can also be seen in Table 6.3.

By performing a multiple correspondence analysis on the alligator data summarised in Table 1.4 using its Burt matrix, the optimal correspondence plot will consist of $S = \min(J, J) - 1 = J - 1$ dimensions since it is a square matrix. Therefore, since $J = 11$, the optimal correspondence plot will consist of no more than $M = 10$ dimensions. Although, since the optimal space from the analysis of the indicator matrix reduces from S to $J - M$ so too does the optimal space when performing multiple correspondence analysis using the Burt matrix. That is, for the simple correspondence analysis of Table 6.3, the optimal correspondence plot consists of $J - M = 11 - 3 = 8$ dimensions.

For the alligator data summarised in Table 1.4, Figure 6.3 gives the two-dimensional correspondence plot from performing multiple correspondence analysis using its Burt matrix. Other than a flipping of the points about the first dimension, its configuration of points looks virtually the same as the configuration of points in the plot obtained from the analysis of the indicator matrix; see Figure 6.1. Therefore, the conclusions reached from observing the general proximity of points from one another in Figure 6.3 are also very similar to those made for Figure 6.1. The three-dimensional correspondence plot, given by Figure 6.4, also shows a similar configuration of points to that of Figure 6.2.

What is different, however, is the weight given to each of the dimensions of the optimal correspondence plot. Performing multiple correspondence analysis by applying simple

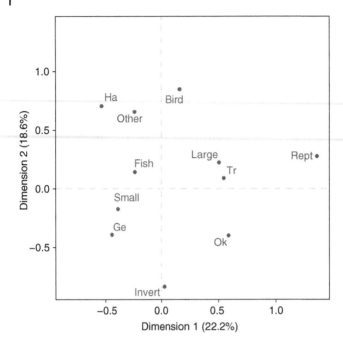

Figure 6.3 Two-dimensional correspondence plot from the multiple correspondence analysis, using the Burt matrix, of Table 1.4.

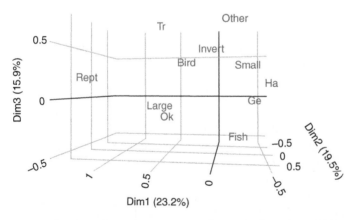

Figure 6.4 Three-dimensional correspondence plot from the multiple correspondence analysis, using the Burt matrix, of Table 1.4.

correspondence analysis to Table 6.3, yields the following squared explained inertia values for the 8-dimensional optimal correspondence plot

$$(\lambda_1^B)^2 = 0.222, \ (\lambda_2^B)^2 = 0.187, \ (\lambda_3^B)^2 = 0.152, \ (\lambda_4^B)^2 = 0.120$$

$$(\lambda_5^B)^2 = 0.100, \ (\lambda_6^B)^2 = 0.082, \ (\lambda_7^B)^2 = 0.065, \ (\lambda_8^B)^2 = 0.029$$

so that the total inertia is

$$\frac{X_B^2}{n} = \sum_{s=1}^{8} (\lambda_s^B)^2$$
$$= 0.222 + 0.187 + \ldots + 0.029$$
$$= 0.957 .$$

An interesting feature of these explained inertia values is that they are linked to those calculated from the indicator matrix through $\lambda_m^B = (\lambda_m^Z)^2$. For example, $\lambda_1^B = \sqrt{0.222} = 0.471 = (\lambda_1^Z)^2$. These explained inertia values are also summarised in Table 6.4 as is the percentage contribution they make to the total inertia and their cumulative percentage contribution of the dimensions. Using these values, we can gauge the quality of the two-(or three-)dimensional correspondence plot when visualising the association between the variables. For example, Table 6.4 shows that the first explained inertia contributes to 23.2% of the total inertia. Therefore, the first dimension of Figure 6.3, and Figure 6.4, visually describes 23.2% of the association between the three variables of Table 1.4. Similarly, the second dimension accounts for 19.5% of this inertia. Therefore, Figure 6.3 visually summarises 42.7% of the association between the variables. It thus shows that Figure 6.3 is a poor quality plot. Despite this, it is an improvement on what the analysis of the indicator matrix provides; recall that Figure 6.1 visually summarises only 33.9% of the association. Adding a third dimension to Figure 6.3, as Figure 6.4 shows, provides a further 15.9% improvement on the quality of the low-dimensional display, and depicts nearly 60% of the association betweeen *Lake*, *Size* and *Food*.

Despite the improvements that analysing the Burt matrix have made when comparing its results with the results from analysing the indicator matrix, further improvements can also be made to the Burt matrix approach to multiple correspondence analysis. One may note that the diagonal tables of the Burt matrix include information that already resides within the off-diagonal tables. This, therefore, inflates the Pearson chi-squared statistic of the Burt matrix by a factor of $J(J - M)/M^2$. Additionally, since the Burt matrix is a super-diagonal

Table 6.4 Inertia values from the correspondence analysis of the Burt matrix of Table 1.4.

Dimension	Explained Inertia	Percentage Inertia	Cumulative Inertia
1	0.222	23.225	23.225
2	0.187	19.513	42.738
3	0.152	15.882	58.621
4	0.120	12.531	71.151
5	0.100	10.424	81.575
6	0.082	8.587	90.162
7	0.065	6.764	96.926
8	0.029	3.074	100.000
Total	0.957	100.000	

matrix, the information contained in the upper diagonal tables is identical to the information contained in the lower diagonal tables. So, to overcome these issues, one can instead perform multiple correspondence analysis by ignoring the duplicated information. Doing so can be done by applying a joint correspondence analysis to the contingency table. The interested reader is invited to peruse the pages of Greenacre (1988, 1990) and Greenacre and Blasius (2006) for a full account, and application, of joint correspondence analysis.

6.4 Stacking

6.4.1 A Definition

Another approach to performing multiple correspondence analysis is via *stacking*. Stacking involves forming a two-way contingency table from the multi-way table by placing each slice (or two-way table) of the array on top of each other. Such an approach has been discussed by, for example, Leclerc (1975), Foucart (1984), Weller and Romney (1990) and Konig (2010). It is not as commonly used in applications as is the indicator or Burt matrices but one may consider it a moderately popular option for performing multiple correspondence analysis.

6.4.2 Stacking and the Alligator Data – *Lake(Size)× Food*

Consider again Agresti's (2007, p. 270) alligator data summarised in Table 1.4. Stacking this contingency table can be done in three ways. For example, one may consider stacking the table so that it reflects an association between the *Size* of the alligator and the *Food* they eat; doing so produces a table of size 10 × 4 where the four columns are the four categories of the *Lake* variable. Alternatively, one may stack each of the *Lake* categories producing a 8 × 4 table; here the 8 reflects the association between the *Size* of the alligators and the *Lake* in which it resides. Such a stacked table is produced as Table 6.5. While one may

Table 6.5 Stacking of *Lake* and *Size* for the Alligator data of Table 1.4.

	Primary Food Choice				
Lake (Size)	Fish	Invert	Rept	Bird	Other
Ha (Small)	23	4	2	2	8
Ha (Large)	7	0	1	3	5
Ok (Small)	5	11	1	0	3
Ok (Large)	13	8	6	1	0
Tr (Small)	5	11	2	1	5
Tr (Large)	8	7	6	3	5
Ge (Small)	16	19	1	2	3
Ge (Large)	17	1	0	1	3

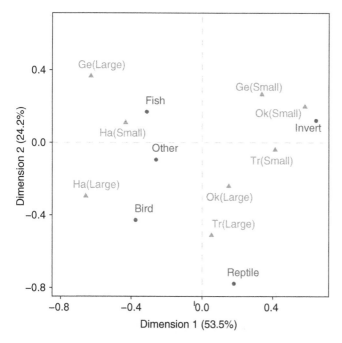

Figure 6.5 Two-dimensional correspondence plot from stacking *Size* and *Lake* from Table 6.5.

examine the association in this manner, it implies that an association exists between the *Lake* and *Size* variables. Alternatively, one may assess this association between the variables by partitioning the three-way chi-squared statistic into three two-way terms (for each pair of variables) and a trivariate term that reflects the overall association structure between all three variables. We do not want to discuss this aspect further but the interested reader may refer to, for example, Beh and Davy (1998, 1999); Goodman (1970); Kateri (2010); Lombardo et al. (2020b) and Beh and Lombardo (2014, Section 11.4.1) for various technical discussions of the issues. Such partitions are sometimes described in the context of log-linear models.

A multiple correspondence analysis on Table 1.4 can be performed by applying a simple correspondence analysis on the stacked table of Table 6.5. Doing so produces the two-dimensional correspondence plot of Figure 6.5. Note that, in this case, the optimal correspondence plot consists of $S = \min (IK, \ J) - 1 = \min (8, \ 5) - 1 = 4$ dimensions. Such a small number of dimensions, when compared with those obtained from the analysis of the indicator and Burt matrices, suggests that Figure 6.5 may well provide a good quality visual summary of the association. We shall determine if this is the case shortly.

Figure 6.5 shows that the *Small* alligators in Lakes *George*, *Oklawaha* and *Trafford* are more likely to feed on *Invertebrates* as their primary source of food than any other food. However, the *Small* alligators in Lake *Hancock* are more likely to have ingested a mix of *Fish* and *Other* material. For the larger sized alligators, their diets are more mixed than their smaller counterparts. For example, *Large* alligators in Lake *Oklawaha* and Lake *Trafford* eat primarily *Reptiles* and *Birds* while *Large* alligators in Lake *Hancock* are more likely to consume *Birds* and *Other* material.

Figure 6.5 also shows that the greatest variation in the eating habits of the different sized alligators are for those from Lake *George*, Lake *Trafford* and Lake *Oklawaha*. This can be seen since the *Large* and *Small* sized alligators from these lakes are located at quite some distance from each other in the plot. On the other hand, it appears that there is relatively little difference in the eating habits of the *Large* and *Small* alligators in Lake *Hancock*. However, there does exist some differences in the eating habits of the two differently sized alligators in Lake *Hancock*. Indeed, by comparing Figure 6.5 with Figure 6.1 (obtained from the analysis of the indicator matrix of Table 1.4) there are also noticeable differences in their configurations. For example, an analysis of the indicator matrix suggests that there is a weak association between *Small* alligators from Lake *Hancock* who eat *Fish* while Figure 6.5 suggests otherwise. Also, Figure 6.1 suggests that the association between *Small* alligators from Lake *Oklawaha* that feed on *Invertibrates* is not as strong as Figure 6.5 shows.

There are three reasons that help account for why Figure 6.5 appears to give differing conclusions to the correspondence plots from the analysis of the indicator and Burt matrices. They are

- By stacking, we are assuming that an association exists between the *Size* of the alligator and the *Lake* in which they are located. Analysing the indicator and Burt matrices make no such assumption.
- Stacking provides a glimpse into the three-way association structure that may exist between the variables of Table 1.4. This is because we are assuming that there exists an association between two of the variables. On the other hand, analysing the indicator and Burt matrices only really considers pair-wise association structures.
- Due to issues concerned with the dimensionality of the optimal correspondence plot. When performing a multiple correspondence analysis on Table 1.4 using the indicator and Burt matrices, the optimal space consisted of 8 dimensions, while stacking produces an optimal space that consists of at most 4 dimensions. Therefore, a two-dimensional correspondence plot from the analysis of the indicator and Burt matrices will generally not provide as good a quality display as stacking does.

So, how good is the two-dimensional correspondence plot of Figure 6.5? We now turn our attention to answering this important question.

By applying a simple correspondence analysis Table 6.5, the four squared-singular values are

$$(\lambda_1^{St})^2 = 0.179, \ (\lambda_2^{St})^2 = 0.081, \ (\lambda_3^{St})^2 = 0.063, \ (\lambda_4^{St})^2 = 0.011$$

so that the total inertia is

$$\frac{X_{St}^2}{n} = \sum_{s=1}^{4} (\lambda_s^{St})^2$$
$$= 0.179 + 0.081 + 0.063 + 0.011$$
$$= 0.334 .$$

We note here that the total inertia of 0.334 is smaller than the total inertia obtained by performing multiple correspondence analysis using the indicator and Burt matrices; recall that their total inertia value was found to be 2.666 and 0.957, respectively. However, since the Pearson chi-squared statistic of Table 6.5 is $X_{St}^2 = 219 \times 0.334 = 73.146$ then, with a p-value that is less than 0.001 at $(8 - 1)(4 - 1) = 21$ degrees of freedom, there is plenty of

Table 6.6 Inertia values from the correspondence analysis of the stacked table of Table 1.4.

Dimension	Explained Inertia	Percentage Inertia	Cumulative Inertia
1	0.179	53.521	53.521
2	0.081	24.231	77.753
3	0.063	18.910	96.661
4	0.011	3.338	100.000
Total	0.334	100.000	

evidence to conclude that a statistically significant association exists between the three variables. This assumes, of course, that an association already exists between the *Lake* and *Size* variables. Although such an assumption is not unreasonable. The two-way contingency table formed from the cross-classification of only the *Lake* and *Size* variables (or, equivalently, by aggregating across the *Food* variable of Table 1.4) results in a Pearson chi-squared statistic of 13.551. With $(2 - 1)(4 - 1) = 3$ degrees of freedom, there exists a statistical significant association between them (p-value = 0.004).

Table 6.6 summarises the four explained inertia values as well as the percentage contribution that each of them makes to the total inertia and their cumulative percentage. We can see then that the first explained inertia accounts for 53.5% of the total inertia which means that the first dimension of Figure 6.5 visually summarises a little more than half of the association between the three variables of Table 1.4. Adding a second dimension summarises a further 24.2% to this association so that the two-dimensional correspondence plot visually summarises about 77.7% of the association. This is very good especially given that analysing the Burt matrix resulted in a two-dimensional plot that summarised only 42.7% of the association. Adding a third dimension to Figure 6.5 (not done here) provides a further 18.9% towards the summary thereby highlighting that a three-dimensional plot visualises nearly all of the association (96.7%). For practitioners of multiple correspondence analysis, such a summary would be very appealing.

6.4.3 Stacking and the Alligator Data – *Food(Size)× Lake*

There are other ways in which we may stack the three variables of Table 1.4. For example, by assuming that there exists an association between the *Size* of the alligator and their primary source of *Food* we can also stack these two variables; doing so produces Table 6.7. Making such an assumption is also not unreasonable; the Pearson chi-squared statistic between these two variables is 14.772 and has a p-value of 0.005 showing that there is evidence of a statistically significant association between them.

The resulting correspondence plot obtained from performing a simple correspondence analysis on Table 6.7 is given by Figure 6.6. It shows that *Small* alligators from Lake *George* are more likely to eat *Invertebrates* while *Large* alligators are more likely to eat *Fish*. The *Large* alligators from Lake *Oklawaha* feed more on *Fish*, *Reptiles* and *Invertebrates*, while

Table 6.7 Stacking of *Size* and primary *Food* source for the alligator data of Table 1.4.

Food (Size)	Lake			
	Hancock	Oklawaha	Trafford	George
Fish (Small)	23	5	5	16
Fish (Large)	7	13	8	17
Invertebrate (Small)	4	11	11	19
Invertebrate (Large)	0	8	7	1
Reptile (Small)	2	1	2	1
Reptile (Large)	1	6	6	0
Bird (Small)	2	0	1	2
Bird (Large)	3	1	3	1
Other (Small)	8	3	5	3
Other (Large)	5	0	5	3

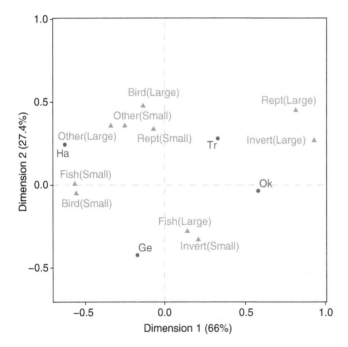

Figure 6.6 Two-dimensional correspondence plot from the multiple correspondence analysis, using the stacking of *Size* and *Primary Food Type*, from Table 6.5.

the alligators of both sizes in Lake *Trafford* appear to have more of a mixed diet than alligators from any of the other lakes. The alligators in Lake *Hancock* appear to be typically smaller than those of the other lakes and also have a varied diet. Note that such findings are consistent with what we described by observing the configuration of points in Figure 6.5. By observing the percentage contribution of the total inertia to each dimension of Figure 6.6, we can see that this two-dimensional correspondence plot provides a visual summary of more than 93% of the association between the three variables of Table 1.4. Figure 6.6 is therefore a very good quality visual display of the association between *Size*, *Lake* and *Food*.

6.5   Final Comments

As we have described at various times throughout this chapter, multiple correspondence analysis does not truly describe the underlying multivariate association structure between the variables of a multi-way contingency table. Instead, since it involves some form of bivariate transformation of the original contingency table, they are at best a way of visualising the various bivariate association structures that exist. In fact, in a section titled "Bivariate or multivariate", de Leeuw (1984) describes this very issue stating, from a principal component analysis perspective (which is equally applicable in the context of multiple correspondence analysis (MCA)),

> One of the things we have learned from the analysis … is that our [MCA], like all classical multivariate analysis, is not strictly *multivariate* but actually *joint bivariate* analysis.

Note here that we have substituted de Leeuw's (1984) reference to "PCA" with "MCA". One may also refer to Lombardo and van Rijckevorsel (2001) for a discussion of this point. Despite these features, multiple correspondence analysis is a very popular method of visually assessing the association between three or more categorical variables. However, it is not the only method by which a visual summary can be made. Rather than transforming a multi-way contingency table into the form of a two-way table, as we have described in this chapter, another way is to treat the data in the *data cube* or *hyper-cube* form that it is in. Therefore, in Chapter 7 we shall examine how correspondence analysis can be formed by preserving this structure. In doing so, we shall be describing *multi-way correspondence analysis*. The foundations of this variant lie in the multi-way extension of the singular value decomposition. There are several such extensions that one can consider, including the PARAFAC/CANDECOMP decomposition. Another extension that we shall describe in Chapter 7 is the Tucker3 decomposition. One may consider Kroonenberg (2008) and Beh and Lombardo (2019) for more information on these methods of decomposition. Like this chapter, our discussion of multi-way correspondence analysis using the Tucker3 decomposition will focus only on nominal categorical variables. However, extensions of the bivariate moment decomposition described in Chapter 2 and hybrid decomposition (see Chapter 5) can be considered. We shall not consider the application of these decomposition methods in the next chapter but, like we described in Part 2, the structure of an ordinal categorical variable can be preserved by scores that are chosen *a-priori*.

7

Multi-way Correspondence Analysis

7.1 An Introduction

Having described in Chapter 6 the correspondence analysis of a multi-way contingency table by first transforming it into the form of a two-way data table, we now focus our attention on treating the data as a *data-cube* or *hyper-cube*. The decision as to which type of "cube" we analyse depends on whether three, or more, variables are being analysed. In the most general case where the purpose of the analysis is to visually summarise the association between many categorical variables the contingency table can be viewed as being in the form of a *hyper-cube*. However, for the sake of simplicity, this chapter will confine the discussion of multi-way correspondence analysis and its application to a three-way contingency table. We shall do so by demonstrating the application of this variant using Agresti's (2002, p. 270) alligator data that is summarised in Table 1.4.

Unlike the approach to correspondence analysis of multiple categorical variables described in Chapter 6, multi-way correspondence analysis does not involve transforming the multi-way contingency table into a two-way form. Instead, the analysis is performed by preserving the multi-way, or *hyper-cube* form of the data. Therefore, for the case of three variables, multi-way correspondence analysis can preserve all pair-wise and three-way association structures between the variables. Recall that this is not a feature of multiple correspondence analysis; see our discussion of this point in Section 6.5.

As we shall describe, since multi-way correspondence analysis preserves the multi-way association structure between the categorical variables, it requires a multi-way generalisation of singular value decomposition (SVD). There are several such generalisations that one may consider. Two such decompositions include parallel factor (or PARAFAC) decomposition and canonical decomposition (CANDECOMP). These methods produce equivalent results and have been extensively described by, for example, Harshman (1970), Harshman and Lundy (1984a,b, 1994), Kiers and Krijnen (1991) and Bro and Kiers (2003).

Another three-way generalisation of SVD is the Tucker3 decomposition (T3D) (Tucker 1966) and has been a topic of discussion by Carlier and Kroonenberg (1996), Kroonenberg (2008) and Kiers et al. (1992). In this chapter, we shall focus our attention on describing and applying the Tucker3 decomposition as a way to perform multi-way correspondence analysis on a three-way contingency table. Further excellent discussions on the various links between these types of decompositions can be found by referring to, for example, Kroonenberg (1989) and Kroonenberg and ten Berge (2011).

An Introduction to Correspondence Analysis, First Edition. Eric J. Beh and Rosaria Lombardo.
© 2021 John Wiley & Sons Ltd. Published 2021 by John Wiley & Sons Ltd.

This chapter will focus on the analysis of three categorical variables. When there exists a symmetric association between all three variables, we refer to the variant of correspondence analysis as *symmetric multi-way correspondence analysis* and this approach involves applying a Tucker3 decomposition to the three-way Pearson residuals, $\gamma_{ijk} - 1$, which we define as a three-way generalisation of Eq. (2.5). When two of the variables are treated as predictor variables and the third variable is treated as being the response variable, the variant of correspondence analysis that we shall describe is referred to as *non-symmetrical multi-way correspondence analysis* and may be considered a three-way extension of *non-symmetrical correspondence analysis* described in Chapter 3. This variant involves applying a Tucker3 decomposition to the Marcotorchino residual, $\pi_{ij|j,k}$ and is a three-way generalisation of Eq. (3.2). We shall be applying both variants to the alligator data summarised in Table 1.4.

7.2 Pearson's Residual $\gamma_{ijk} - 1$ and the Partition of X^2

7.2.1 The Pearson Residual

In Section 2.3.1 we described that the foundations of correspondence analysis can be viewed by assessing departures from independence using the Pearson ratio, γ_{ij}, defined by Eq. (2.4). We can extend this idea for the analysis of three symmetrically associated variables. For example, when testing whether there exists a statistically significant association between three categorical variables, complete independence in the (i, j, k)th cell arises when

$$p_{ijk} = p_{i\bullet\bullet}p_{\bullet j\bullet}p_{\bullet\bullet k}$$

for all $i = 1, 2, \ldots, I, j = 1, 2, \ldots, J$ and $k = 1, 2, \ldots, K$. The departure from independence of this cell can be assessed by specifying that

$$p_{ijk} = \gamma_{ijk}p_{i\bullet\bullet}p_{\bullet j\bullet}p_{\bullet\bullet k}$$

where

$$\gamma_{ijk} = \frac{p_{ijk}}{p_{i\bullet\bullet}p_{\bullet j\bullet}p_{\bullet\bullet k}}$$

is the *Pearson ratio* of the cell. Therefore, variations of γ_{ijk} from 1 imply that, for the (i, j, k)th cell of the table, there exists a departure from independence. Note that complete independence will arise in the three-way contingency table if all

$$\gamma_{ijk} = 1$$

for all $i = 1, 2, \ldots, I, j = 1, 2, \ldots, J$ and $k = 1, 2, \ldots, K$.

A related measure that we shall be considering in this chapter for three symmetrically associated categorical variables is the Pearson residual. For the (i, j, k)th cell, this residual is defined as

$$\gamma_{ijk} - 1 = \frac{p_{ijk}}{p_{i\bullet\bullet}p_{\bullet j\bullet}p_{\bullet\bullet k}} - 1 \ . \tag{7.1}$$

Therefore, Pearson's chi-squared statistic defined by Eq. (1.6) can be expressed in terms of this residual such that

$$X^2 = n\sum_{i=1}^{I}\sum_{j=1}^{J}\sum_{k=1}^{K}p_{i\bullet\bullet}p_{\bullet j\bullet}p_{\bullet\bullet k}(\gamma_{ijk} - 1)^2 \ .$$

So, if $\gamma_{ijk} - 1 = 0$ for all $i = 1, 2, \ldots, I, j = 1, 2, \ldots, J$ and $k = 1, 2, \ldots, K$ then there is complete independence between the three variables so that $X^2 = 0$.

7.2.2 The Partition of X^2

When there exists a statistically significant association between the variables of a three-way contingency table, further information can be obtained on the structure of this association by partitioning X^2 into four terms such that

$$X^2 = X_{IJ}^2 + X_{IK}^2 + X_{JK}^2 + X_{IJK}^2 .$$

Here

- X_{IJ}^2 is the Pearson chi-squared statistic for the cross-classification of the row and column variables. The statistical significance of this term may be determined by comparing it with the $1 - \alpha$ percentile of the chi-squared distribution with $(I - 1)(J - 1)$ degrees of freedom.
- X_{IK}^2 is the Pearson chi-squared statistic for the cross-classification of the row and tube variables. The statistical significance of this term may be determined by comparing it with the $1 - \alpha$ percentile of the chi-squared distribution with $(I - 1)(K - 1)$ degrees of freedom.
- X_{JK}^2 is the Pearson chi-squared statistic for the cross-classification of the column and tube variables. The statistical significance of this term may be determined by comparing it with the $1 - \alpha$ percentile of the chi-squared distribution with $(J - 1)(K - 1)$ degrees of freedom.
- X_{IJK}^2 is the chi-squared statistic that assesses the three-way association between the all three categorical variables of the contingency table. The statistical significance of this term may be determined by comparing it with the $1 - \alpha$ percentile of the chi-squared distribution with $(I - 1)(J - 1)(K - 1)$ degrees of freedom.

See, for example, Lancaster (1951), Loisel and Takane (2016) and Lombardo et al. (2020b) for a description of this partition for three nominal categorical variables. When at least one of the variables consist of ordered categories such a partition was discussed by Beh and Davy (1998, 1999).

7.2.3 Partition of X^2 for the Alligator Data

Consider again the alligator data summarised in Table 1.4. In Section 1.6.5 we established that, with a Pearson chi-squared statistic of $X^2 = 85.492$ and 31 degrees of freedom, there is a statistically significant association between the *Size*, *Food* and *Lake* variables of the contingency table. However, this statistic does not tell us anything about whether some, or all, pairs of variables are associated. Nor does the statistic tell us if their exists a truly three-way association between these variables. To find out, we can partition X^2 into four chi-squared terms that tell us more about the association structure. Table 7.1 summarises the terms of this partition, their percentage contribution (%) to X^2, degrees of freedom (df) and p-value.

Table 7.1 shows that there is a statistically significant association between each pair, and all three, variables of Table 1.4. It also shows those associations that are the most dominant. For example, Table 7.1 shows the following association features:

Table 7.1 Partition of X^2 for Table 1.4

	X_{IJ}^2	X_{IK}^2	X_{JK}^2	X_{IJK}^2	X^2
Values	14.772	13.551	37.729	19.440	85.492
%	17.279	15.850	44.131	22.739	100.000
df	4	3	12	12	31
p-value	0.005	0.004	< 0.001	0.078	< 0.001

- The association between *Size* and *Food*, quantified by the Pearson chi-squared statistic $X_{IJ}^2 = 14.772$, contributes to about 17.3% of the complete association structure between the three variables. Therefore, there is a statistically significant association between the *Size* of an alligator and the type of *Food* it eats.
- The association between *Size* and *Lake*, quantified by the statistic $X_{IK}^2 = 13.551$, accounts for about 15.9% of the total association that exists between the variables. Therefore, there is a statistically significant association between the *Size* of an alligator and the *Lake* in which it lives.
- The most dominant source of association in Table 1.4 is the association between the *Food* and *Lake* variables. The Pearson chi-squared statistic for this association is $X_{JK}^2 = 37.729$ and accounts for nearly half (44.1%) of the association that exists between the three variables of Table 1.4. Therefore, there is a statistically significant, and very dominant, association between the type of *Food* an alligator eats and the *Lake* in which it lives.
- If one were to adopt a 0.10 level of significance then there exists a statistically significant association between all three variables of Table 1.4. The chi-squared term here is $X_{IJK}^2 = 19.440$ and accounts for 22.7% of the total association that exists in the contingency table. While this term contributes more to X^2 than X_{IJ}^2 and X_{IK}^2, it has more degrees of freedom to work with which is the reason for its relatively large p-value.

If all three categorical variables consisted of ordered categories, these four terms can be further partitioned to reveal far more detail about the underlying structure of the association. For the sake of simplicity, we shall not consider this issue further here. However, the interested reader is invited to refer to Beh and Davy (1998, 1999), Lombardo et al. (2016b), Lombardo and Beh (2017) and Lombardo et al. (2020a) for more information on this issue.

While we may now know from the partition of X^2 what variables of Table 1.4 have a statistically significant association, its terms still tell us nothing about HOW they are associated, only that they are. That is, we still do not know how exactly the categories of one variable interact (or not) with categories of another variable. While Chapter 6 explored the application of correspondence analysis to Table 1.4, the way in which it did so was to transform the three-way contingency table into a two-way tabular form. Therefore the three-way association structure is reduced to investigating multiple pair-wise association structures. Rather than performing correspondence analysis in this manner we can preserve the three-way structure of the contingency table and perform multi-way correspondence analysis instead. We now introduce the multi-way correspondence analysis of symmetrically associated variables.

7.3   Symmetric Multi-way Correspondence Analysis

7.3.1   Tucker3 Decomposition of $\gamma_{ijk} - 1$

Suppose we have three nominal categorical variables that are symmetrically associated and are cross-classified to form a three-way contingency table. In Chapter 6 we described that one way to perform a correspondence analysis on this table is to transform it into a two-way table and then apply a SVD to this transformed table. Another way is to preserve the inherent three-way association structure in the contingency table by treating it as a *data cube*. However, in doing so we can no longer use SVD. Instead we must make use of a three-way generalisation of SVD. One such generalisation is the Tucker3 decomposition; from here on we shall abbreviate it to T3D. Therefore, a correspondence analysis of a three-way contingency table may be undertaken by applying a T3D to the three-way Pearson residuals, $\gamma_{ijk} - 1$, such that

$$\gamma_{ijk} - 1 = \sum_{p=1}^{P}\sum_{q=1}^{Q}\sum_{r=1}^{R}\lambda_{pqr}a_{ip}b_{jq}c_{kr} + e_{ijk} \tag{7.2}$$

where P, Q and R are the fixed number of components necessary for the visualisation of the rows, columns and tubes, respectively. Here, a_{ip} is the score of the ith row along the pth dimension of a correspondence plot, for $i = 1, 2, \ldots, I$ and $p = 1, 2, \ldots, P \le I$. These scores are constrained so that

$$\sum_{i=1}^{I}p_{i\bullet\bullet}a_{ip} = 0, \quad \sum_{i=1}^{I}p_{i\bullet\bullet}a_{ip}^2 = 1$$

for each of the P components. Similarly, the score for the jth column along the qth dimension of the correspondence plot is defined as b_{jq} for $j = 1, 2, \ldots, J$ and $q = 1, 2, \ldots, Q \le J$, and are constrained so that

$$\sum_{j=1}^{J}p_{\bullet j\bullet}b_{jq} = 0, \quad \sum_{j=1}^{J}p_{\bullet j\bullet}b_{jq}^2 = 1$$

for each of the Q column components. Finally, c_{kr} is the score of the kth tube category along the rth dimension for $j = 1, 2, \ldots, J$ and $r = 1, 2, \ldots, R \le K$. These scores are constrained so that

$$\sum_{k=1}^{K}p_{\bullet\bullet k}c_{kr} = 0, \quad \sum_{k=1}^{K}p_{\bullet\bullet k}c_{kr}^2 = 1$$

for each of the R tube components.

These constraints imply that, from a statistical perspective, the expectation and variance of a_{ip}, say, along the pth dimension of the resulting visual summary is

$$\mathrm{E}(a_{ip}) = \sum_{i=1}^{I}p_{i\bullet\bullet}a_{ip}$$
$$= 0$$

and

$$\text{Var}(a_{ip}) = E(a_{ip}^2) - [E(a_{ip})]^2$$
$$= \sum_{i=1}^{I} p_{i\bullet} a_{ip}^2 - 0$$
$$= 1,$$

respectively.

The λ_{pqr} term in Eq. (7.2) may be viewed as a three-way generalisation of a singular value and is referred to as the (p, q, r)th element of the *core array*. These terms reflect a specific feature of the association that exist between the three variables of a contingency table; see Carroll and Chang (1970), Kroonenberg (1983, 1994) and Kroonenberg and de Leeuw (1980). This structure may be viewed as a three-way extension of the correlation term defined by Eq. (2.10) (for nominal correspondence analysis) and Eq. (4.8) (for ordered correspondence analysis) since

$$\lambda_{pqr} = \sum_{i=1}^{I}\sum_{j=1}^{J}\sum_{k=1}^{K} p_{i\bullet\bullet}p_{\bullet j\bullet}p_{\bullet\bullet k}(\gamma_{ijk} - 1)a_{ip}b_{jq}c_{kr}$$
$$= \sum_{i=1}^{I}\sum_{j=1}^{J}\sum_{k=1}^{K} p_{ijk}a_{ip}b_{jq}c_{kr}. \tag{7.3}$$

Such a result may also be derived by noting that

$$\lambda_{pqr} = \text{GenCorr}(a_{ip}, b_{jq}, c_{kr})$$
$$= \sum_{i=1}^{I}\sum_{j=1}^{J}\sum_{k=1}^{K} p_{ijk}\frac{(a_{ip} - E(a_{ip}))}{\sqrt{\text{Var}(a_{ip})}}\frac{(b_{jq} - E(b_{jq}))}{\sqrt{\text{Var}(b_{jq})}}\frac{(c_{kr} - E(c_{kr}))}{\sqrt{\text{Var}(c_{kr})}}$$
$$= \sum_{i=1}^{I}\sum_{j=1}^{J}\sum_{k=1}^{K} p_{ijk}a_{ip}b_{jq}c_{kr}. \tag{7.4}$$

Here GenCorr (a_{ip}, b_{jq}, c_{kr}) is the generalised correlation between a_{ip}, b_{jq}, and c_{kr} and is analogous to the generalised correlation described by Rayner and Beh (2009) for the analysis of three ordered categorical variables.

Negative values of λ_{pqr} reflect a negative association that exists between the components while a positive value indicates that there exists a positive association. If there is complete independence between the three variables so that $p_{ijk} = p_{i\bullet\bullet}p_{\bullet j\bullet}p_{\bullet\bullet k}$ then

$$\lambda_{pqr} = \sum_{i=1}^{I}\sum_{j=1}^{J}\sum_{k=1}^{K} p_{i\bullet\bullet}p_{\bullet j\bullet}p_{\bullet\bullet k}a_{ip}b_{jq}c_{kr}$$
$$= \left(\sum_{i=1}^{I} p_{i\bullet\bullet}a_{ip}\right)\left(\sum_{j=1}^{J} p_{\bullet j\bullet}b_{jq}\right)\left(\sum_{k=1}^{K} p_{\bullet\bullet k}c_{kr}\right)$$
$$= 0.$$

This result also verifies a point raised by Kroonenberg (2008, p. 55) on these value. He noted that if $\lambda_{pqr} = 0$ then the combination of the pth row, qth column and rth tube components does not contribute to the association structure.

Lastly, the term e_{ijk} is the error between the observed value of $\gamma_{ijk} - 1$ and its reconstituted value using Eq. (7.2), for some value of P, Q and R. T3D guarantees that with the full set of components, such that $P = I$, $Q = J$ and $R = K$, the $\gamma_{ijk} - 1$ values will be perfectly reconstituted so that all $e_{ijk} = 0$ for $i = 1, \ldots, I, j = 1, \ldots, J$ and $k = 1, \ldots, K$. Therefore, if $P < I$, $Q < J$ and/or $R < K$ then an error will be introduced into the reconstitution and so we get the following approximation of the (i, j, k)th Pearson residual

$$\hat{\gamma}_{ijk} - 1 = \sum_{p=1}^{P}\sum_{q=1}^{Q}\sum_{r=1}^{R} \lambda_{pqr} a_{ip} b_{jq} c_{kr} \; .$$

Therefore, to determine the "best" choice of P, Q and R and, in doing so, calculate their associated values of a_{ip}, b_{jq}, c_{kr} and λ_{pqr}, one can minimise the weighted sum-of-squares of the e_{ijk} terms. That is, their solution can be found by minimising

$$\text{SSE} = \sum_{i=1}^{I}\sum_{j=1}^{J}\sum_{k=1}^{K} p_{i\bullet\bullet} p_{\bullet j\bullet} p_{\bullet\bullet k} e_{ijk}^{2}$$

$$= \sum_{i=1}^{I}\sum_{j=1}^{J}\sum_{k=1}^{K} p_{i\bullet\bullet} p_{\bullet j\bullet} p_{\bullet\bullet k} (\gamma_{ijk} - \hat{\gamma}_{ijk})^{2}$$

using, typically, a weighted alternating least-squares algorithm.

The T3D of a three-way array has been extensively discussed in the correspondence analysis, and related, literature since it was first put forward by Tucker (1963) and elaborated upon by Tucker (1964, 1966). More recent discussions of the T3D can also be found in Kiers and Krijnen (1991), Kiers et al. (1992) and Kroonenberg (1983, 2008). Further extensions and applications of this decomposition approach can also be seen by referring to Kroonenberg (1985, 1987, 1989), Smilde (1992), Kiers (2000), Smilde et al. (2004a), Smilde et al. (2004b), Rocci and Vichi (2005), Bro (2006), Ellis et al. (2006), Spain et al. (2010), Gallo and Buccianti (2013) and Lombardo et al. (2019). Applications of the T3D for the correspondence analysis of a three-way contingency table was first proposed by Carlier and Kroonenberg (1996) and elaborated upon by van Herk and van de Velden (2007) and Beh and Lombardo (2014). However, such an approach to correspondence analysis has not been implemented into any commercially available statistics packages, although one may undertake such an analysis using the R package CA3variants; see Lombardo and Beh (2016) for more information on this issue.

One may note that Eq. (7.2) appears to be akin to what would be obtained by extending to the three-way case the BMD of $\gamma_{ij} - 1$ given by Eq. (4.7). However, there are some differences. While both expressions involve the summation across the full complement of their respective components there is no requirement that the components from the T3D have polynomial order restrictions. That is, it is not necessary that the a_{i1} terms, say, are linearly arranged or that a_{i2} have a quadratic set of values. Also, BMD involves summing across the number of categories minus one for each variable, while P, Q and R are selected to achieve the minimum SSE. Nor is there any requirement that the λ_{pqr} are arranged in any particular order. However, Pearson's chi-squared statistic of a three-way contingency table can be expressed in terms of them such that

$$X^{2} = n\sum_{p=1}^{P}\sum_{q=1}^{Q}\sum_{r=1}^{R} \lambda_{pqr}^{2}$$

when $P = I, Q = J$ and $R = K$ so that X^2/n is the total inertia of the contingency table. When $P < I, Q < J$ and/or $R < K$, this X^2 value is an approximation of the true chi-squared statistic and we therefore can express how similar, or different, this approximation is in terms of a percentage. Like we have done in the earlier chapters, a solution to a_{ip}, b_{jq} and c_{kr} that leads to an approximation of X^2 that is at least 70% of the total inertia will be considered a good solution and the resulting correspondence plot will be deemed a good visual summary of the association.

7.3.2 T3D and the Analysis of Two Variables

As we demonstrated in Section 7.2, Pearson's chi-squared statistic for a three-way contingency table consisting of symmetrically associated variables can be partitioned into four terms; three of these terms reflect the three combinations of pair-wise association while the fourth term reflected the three-way association between the variables. This partition can also be applied to the T3D of $\gamma_{ijk} - 1$ and such a partition of this Pearson residual was described in Carlier and Kroonenberg (1996). For example, suppose we wish to investigate the association between the row and column variables only. Then this can be done using the following measure

$$\gamma_{ij\bullet} - 1 = \sum_{k=1}^{K} p_{\bullet\bullet k}(\gamma_{ijk} - 1)$$

$$= \sum_{k=1}^{K} p_{\bullet\bullet k}\left(\frac{p_{ijk}}{p_{i\bullet\bullet}p_{\bullet j\bullet}p_{\bullet\bullet k}} - 1\right)$$

$$= \frac{p_{ij\bullet}}{p_{i\bullet\bullet}p_{\bullet j\bullet}} - 1$$

which is just the Pearson residual of the (i, j)th cell of the two-way contingency table obtained by aggregating across each of the K tubes. Therefore, the features obtained from the T3D of $\gamma_{ijk} - 1$ can also be used for the bivariate partition so that

$$\gamma_{ij\bullet} - 1 = \sum_{p=1}^{P}\sum_{q=1}^{P} a_{ip}b_{jp}\lambda_{pq\bullet} + e_{ij\bullet} \tag{7.5}$$

where

$$\lambda_{pq\bullet} = \sum_{i=1}^{I}\sum_{j=1}^{J} p_{ij\bullet}a_{ip}b_{jq} \tag{7.6}$$

so that

$$X_{IJ}^2 = \sum_{p=1}^{P}\sum_{q=1}^{Q} \lambda_{pq\bullet}^2 ,$$

for $P = I$ and $Q = J$. Note, that Eq. (7.5) is mathematically analogous to Eq. (4.7), the only difference between the two being that Eq. (7.5) treats the row and column variables as consisting of nominal categories while Eq. (4.7) treats them as consisting of ordered categories. Thus, $\lambda_{pq\bullet}$ has a similar mathematical structure to λ_{uv}; see Eq. (4.8) but these two generalised correlations have different interpretations. Despite these differences, the sum-of-squares of λ_{uv} and $\lambda_{pq\bullet}$ are directly related to Pearson's chi-squared statistic of the contingency table formed from the cross-classification of I row and J column categories.

7.3.3 On the Choice of the Number of Components

Unlike SVD or BMD, the choice of P, Q and R from the T3D of $\gamma_{ijk} - 1$ is not completely arbitrary. This is because, by using the least-squares algorithm to determine a_{ip}, b_{jq}, c_{kr} and λ_{pqr}, changing P, Q and R will change their value. One can start with setting $P = I$, $Q = J$ and $Q = K$ and obtain the optimal solution (since $SSE = 0$) so that 100% of the total inertia is accounted for. Depending on the size of the contingency table being analysed, if I, J and K are large this can create convergences issues since the algorithm is having to calculate, and recalculate (iteratively) values for these four terms multiple times. This makes choosing P, Q and R difficult. Ideally then, one would set $P < I$, $Q < J$ and $R < K$ but this choice cannot be arbitrary since there are restrictions put in place on what P, Q and R can be to ensure convergence of the algorithm arises; these restrictions are that $PQ \geq R$, $QR \geq P$ and $RP \geq Q$.

So, to determine the best possible number of row, column and tube components various strategies have been put forward. For example, Kroonenberg and Oort (2003) and Murakami and Kroonenberg (2003) considered selecting P, Q and R based on the change in the shape of a type of scree-plot while Ceulemans and Kiers (2006) adopted a similar strategy but used a convex hull approach to determine the change. Kiers (1997, 1998), Ten Berge and Kiers (1999), Andersson and Henrion (1999), Rocci (2001) and Kroonenberg (2005) suggested rotating the solutions to a_{ip}, b_{jq}, c_{kr} and λ_{pqr} while Rocci (1992), Kiers et al. (1997) and Kiers (1998) considered constraining some of the elements in the core array to zero. Another strategy is through the "core consistency diagnostic" of Bro and Kiers (2003) (although they focused their attention on the PARAFAC, not the Tucker3, decomposition) while a more straightforward approach of focusing only on those components that explain as much of the total inertia as reasonably possible was suggested by Timmerman and Kiers (2000). We invite the interested reader to investigate each of these strategies at their leisure. For our analysis of the alligator data summarised in Table 1.4, we will adopt the most practical criterion. That is, choosing P, Q and R that provides computational simplicity and explains as much as is reasonable of the total inertia. Section 7.3.4 will investigate this issue while demonstrating the application of the T3D to the $\gamma_{ijk} - 1$ values of Table 1.4.

7.3.4 Tucker3 Decomposition of $\gamma_{ijk} - 1$ and the Alligator Data

To demonstrate the features of T3D, consider again the alligator data summarised in Table 1.4. The numerical features of this decomposition are given as follows. Table 7.2 summarises, five combinations of P, Q and R, the explained inertia value, the percentage of

Table 7.2 Summary of inertia values for five combinations of P, Q and R from the T3D of $\gamma_{ijk} - 1$ for Table 1.4

P	Q	R	Inertia	% Explained Inertia	Iterations
1	1	1	0.126	32.22%	27
2	1	1	0.124	31.65%	22
2	2	2	0.261	66.82%	9
2	3	2	0.298	76.30%	16
2	5	4	0.390	100.00%	1

X^2/n that it explains and the number of iterations required for the least-squares procedure to converge to three decimal places. It shows that if only a single component was used to discriminate between the row, column and tube categories, only 32% of the total inertia would be explained. This is far too small a value to produce a meaningful numerical, or visual, assessment of the association between the three variables. As P, Q and R approach $P = I = 2$, $Q = J = 5$ and $R = K = 4$ we find that the percentage of the total inertia that is explained increases to 100%, i.e. it is equal to the total inertia, X^2/n, of the three-way contingency table.

We can also see that when $P = Q = R = 2$, 66.82% of the total inertia is reflected when using the first two row, column and tube components. This is close to the 70% criteria we adopted in Chapter 2 for deeming a correspondence plot to be a good visual summary of the association between the three variables of Table 1.4. While it does not quite meet the 70% threshold, choosing $P = Q = R = 2$ does provide advantages for simply summarising the association between the three variables using a two-dimensional correspondence plot. Therefore, setting $P = Q = R = 2$ will determine the values of a_{ip}, b_{jq}, c_{kr} and λ_{pqr} and these can be calculated using the R package CA3variants. Doing so, we find that the row components from the T3D are summarised in Table 7.3. Similarly, Table 7.4 and Table 7.5 summarise the column and tube components, respectively, from this decomposition. Note that 76.30% of the total inertia is explained if we were to consider $P = 2$, $Q = 3$ and $R = 2$, however we prefer the more parsimonious number of components $P = 2$, $Q = 2$ and $R = 2$.

The values in these tables may be viewed in the same way as the elements of singular vectors from the GSVD of $\gamma_{ij} - 1$; see our discussion in Section 2.4.3. That is, if one were to construct a two-dimensional visual summary of the association between the three variables, (a_{i1}, a_{i2}) is the standard coordinate of the ith row category. Similarly, (b_{j1}, b_{j2}) and (c_{k1}, c_{k2}) is the standard coordinate of the jth column and kth tube category, respectively. This would therefore mean that, from the values summarised in Table 7.3, the location of the point for the *Small* alligators would be located on the left of the display while the location of the

Table 7.3 The a_{ip} values from the T3D of $\gamma_{ijk} - 1$ for Table 1.4

	a_{i1}	a_{i2}
Small	−0.587	0.809
Large	0.809	0.587

Table 7.4 The b_{jq} values from the T3D of $\gamma_{ijk} - 1$ for Table 1.4

	b_{j1}	b_{j2}
Fish	−0.427	−0.650
Invertebrate	0.799	−0.057
Reptile	0.318	−0.749
Bird	−0.118	−0.112
Other	−0.252	0.031

Table 7.5 The c_{kr} values from the T3D of $\gamma_{ijk} - 1$ for Table 1.4

	c_{k1}	c_{k2}
Hancock	−0.578	−0.014
Oklawaha	0.670	−0.059
Trafford	0.466	0.067
George	0.001	0.996

Large alligators would be on the right, thereby showing a clear difference between the two sizes of alligators.

Table 7.4 shows that of the variety of different *Food* items that are consumed, the first component is dominated by alligators that eat *Invertebrate* animals as their primary source of food. Only those that eat *Reptiles* have a similar eating habit (at least along the first component) while the remaining *Food* items would be located on the left of a visual display. However, if a two-dimensional depiction of these categories using their standard coordinates were to be constructed it would show that the alligators eating habits of the five *Food* items are very different. Certainly, while *Fish* and *Reptiles* have a very similar position along the second component they are very different in terms of the two *Size* categories.

Suppose we now examine the standard coordinate (c_{k1}, c_{k2}) for $k = 1, 2, 3, 4$ that represent the position of the four *Lake* categories in a two-dimensional display. Table 7.5 clearly shows that the size and eating habits of alligators in *Hancock* are very different from those in other lakes. If we were to confine our attention to just the first component, this would suggest that alligators in Lake *Oklawaha* and Lake *Trafford* are very similar in size and in what they eat. However there is some variation in their second component values.

As we have described throughout this book, using standard coordinates does not produce an appropriate visual summary of the association since the (generalised) correlation between the three variables has not been included in the definition of the coordinates. That is, their definition does not incorporate the sign or magnitude of the elements of the core array, λ_{pqr}. Table 7.6 summarises magnitude and sign of the $\sqrt{n}\lambda_{pqr}$ terms and shows that

Table 7.6 The core array elements, $\sqrt{n}\lambda_{pqr}$ from the T3D of $\gamma_{ijk} - 1$ of Table 1.4

	c_{k1}	
	b_{j1}	b_{j2}
a_{i1}	0.126	−4.764
a_{i2}	4.599	0.088

	c_{k2}	
	b_{j1}	b_{j2}
a_{i1}	−3.570	−0.047
a_{i2}	0.035	0.713

there is a very strong positive association between the second row component, first column component and first tube component. That is, the generalised correlation associated with the interaction $a_{i2}b_{j1}c_{k1}$ is $\lambda_{211} = 4.599$. Further evidence of a strong association, albeit a negative one, between the three variables is evident for

- The first row component, second column component and first tube component. That is, the generalised correlation associated with the interaction $a_{i1}b_{j2}c_{k1}$ is $\lambda_{121} = -4.764$.
- The first row component, first column component and second tube component. That is, the generalised associated with the interaction $a_{i1}b_{j1}c_{k2}$ is $\lambda_{112} = -3.570$.

Therefore, there is clear evidence of a strong three-way association existing between the variables. In Section 7.4 we show how the values summarised in Table 7.3, Table 7.4, Table 7.5 and Table 7.6 can be used to construct a low-dimensional visual summary of the association between the three variables of Table 1.4. We can also verify that setting $P = Q = R = 2$ leads to 67% of the total inertia being explained since

$$\sum_{p=1}^{2}\sum_{q=1}^{2}\sum_{r=1}^{2}(\sqrt{n}\lambda_{pqr})^2 = (0.126)^2 + (-4.764)^2 + \ldots + (0.713)^2$$

$$= 57.127 .$$

So, $100 \times 57.127/85.492 = 66.82\%$ of the total inertia is explained by setting $P = Q = R = 2$; exactly the same proportion of the inertia that is summarised in Table 7.6 for this P, Q and R.

While an optimal solution is obtained when $P = 2$, $Q = 5$ and $R = 4$, the inertia obtained by just selecting the first two row, column and tube components from this optimal analysis does not always produce the highest percentage of explained inertia. Nor is there any guarantee that the amount of inertia from doing so will be any better than when using those values of λ_{pqr} obtained by setting $P = Q = R = 2$. To show this, Table 7.7 summarises the core array elements, λ_{pqr} for $p = q = r = 2$, when $P = 2$, $Q = 5$ and $R = 4$. While the chi-squared statistic of Table 1.4 is 85.492, the sum-of-squares of the values in Table 7.7 gives an inertia value of

$$(1.410)^2 + (4.444)^2 + \ldots + (0.970)^2 = 53.431,$$

Table 7.7 The first two sets of optimal $\sqrt{n}\lambda_{pqr}$ from the T3D of $\gamma_{ijk} - 1$ of Table 1.4

	c_{k1}	
	b_{j1}	b_{j2}
a_{i1}	1.410	4.444
a_{i2}	4.346	-0.974

	c_{k2}	
	b_{j1}	b_{j2}
a_{i1}	2.870	0.574
a_{i2}	-1.533	0.970

and accounts for $100 \times 53.431/85.492 = 62.50\%$ of the total inertia. This figure is less than the percentage of the explained inertia obtained by calculating the sum-of-squares of the $\sqrt{n}\lambda_{pqr}$ values when $P = Q = R = 2$; see Table 7.6. Furthermore, recall that this inertia, and its explained inertia, is summarised in Table 7.2 and is 0.261 and 66.82%, respectively. There-fore, unlike the classical approach to simple correspondence analysis, and as tempting as it might be (for the sake of simplicity), multi-way correspondence analysis does not allow the analyst to merely select the first two components, say, and their core array elements, for a visual summary of the association. One must first determine an appropriate value of P, Q and R (as we have done) that meets the criterion necessary for constructing what may be considered a reasonable plot of the association.

We now turn our attention to how a visual summary of the association between the row, column and tube variables can be obtained for our given values of P, Q and R.

7.4 Constructing a Low-Dimensional Display

7.4.1 Principal Coordinates

In Section 7.3.4 we described that one could treat a_{ip}, b_{jq} and c_{kr} as elements of the row, column and tube standard coordinates. However, such coordinates do not reflect the nature of the association structure that is captured in the λ_{pqr} values. Instead one could construct principal coordinates in a similar manner to those derived for a doubly ordered contingency table – recall that the row and column principal coordinates for such a contingency table were defined by Eq. (4.12) and Eq. (4.13). That is, in the context of a three-way contingency table such coordinates are defined as

$$f_{iqr} = \sum_{p=1}^{P} a_{ip} \lambda_{pqr} \tag{7.7}$$

$$g_{ipr} = \sum_{q=1}^{Q} b_{jq} \lambda_{pqr} \tag{7.8}$$

$$h_{ipq} = \sum_{r=1}^{R} c_{kr} \lambda_{pqr} \; . \tag{7.9}$$

The problem though is that these coordinates imply that a single row, column or tube point must be depicted along a dimension that reflects the interaction between two com-ponents. For such an introductory discussion as we are making here this idea is beyond the scope of this book. This is because additional complexities are introduced into the inter-pretation of the correspondence plot when using these coordinates to visually summarise the association between the three variables. Instead, we construct a biplot to visualise the association.

Once we go beyond the analysis of two categorical variables, there are many types of biplot that one can construct. For three variables the biplot that we focus on in Section 7.4.2 is the *interactive biplot*. Lombardo et al. (2016b) described such a biplot for the analysis of ordered asymmetrically associated categorical variables while Kroonenberg (2008, Section 11.5) gives an account of various different types of biplot for multi-way data analysis. These include the *nested-mode biplot* which is analogous in structure to the interactive biplot.

7.4.2 The Interactive Biplot

Two Types of Interactive Biplot

To visually summarise the symmetrical association between the variables of a three-way contingency table, we can construct an interactive biplot. Such a biplot is also referred to as a *nested-mode biplot* and has been discussed at some length by Carlier and Kroonenberg (1996), Lombardo et al. (1996) and Beh and Lombardo (2014). Kroonenberg (2008, Section 11.5) also discussed this biplot and other variations in his description of joint displays for multi-way data analysis. As we shall see shortly, such a visual summary provides the analyst with a single representation of the association between the three variables. The interested reader is also directed to Gabriel (1971), Greenacre (2010, 2017), Gower et al. (2011 and Gower et al. (2014, 2016) for a discussion of the various biplots, and issues, that are now available.

When constructing an interactive biplot the key word to keep in mind is *interactive*. As the name suggests the biplot is formed in such a way that it reflects a particular interaction between the categories from any two variables. The implication then is that this type of biplot is constructed by assuming, or because it is known, that a statistically significant association exists between those two variables. This can happen in two ways:

- A *column–tube interactive biplot*, say, defines a set of principal coordinates for the column and tube variables assuming that an interaction exists between them. Therefore, a single point is used to represent how a particular pair of column–tube categories impacts the association in the contingency table and its projection from the origin reflects how closely it is associated with the row categories. So, for this biplot, a row category is depicted as a projection from the origin to its position in a biplot using its standard coordinate while the interaction between a pair of column-tube categories is depicted using a single point at the position determined by its principal coordinate. Constructing a biplot in this manner means that the distance of a column–tube point from the row points projection can be meaningfully interpreted. The resulting visual display is therefore a form of *isometric biplot*.
- A *row interactive biplot* depicts a row point as a single point defined by its principal coordinate while the interaction between a pair of column and tube categories are depicted as a projection from the origin to its position defined by its standard coordinate. Therefore, the distance of a row principal coordinate from the projection of the column–tube standard coordinate assesses how strong the interaction is between a row category and a column–tube category pair. Note that such an interactive biplot is also an example of an *isometric biplot*. For more information on this type of biplot, one may refer to Carlier and Kroonenberg (1996), Beh and Lombardo (2014) and Lombardo and Beh (2017).

We now turn our attention to defining the coordinate system for these two interactive biplots.

Column–Tube Interactive Biplot

Suppose we wish to construct a *column–tube interactive biplot*. In doing so, the *i*th row standard coordinate and the (j, k)th column–tube principal coordinate along the *p*th dimension of the biplot is defined by

$$\tilde{f}_{ip} = a_{ip} \tag{7.10}$$

$$\tilde{g}_{jkp} = \sum_{q=1}^{Q}\sum_{r=1}^{R} b_{jq}\lambda_{pqr}c_{kr}\,, \tag{7.11}$$

respectively, for $p = 1, 2, \ldots, P$. Therefore, this biplot is an optimal low-dimensional space that consists of P dimensions.

Note that Eq. (7.11) implies that an association exists between the column and tube variables. If there exists complete independence between these two variables, i.e. $p_{ijk} = p_{ij\bullet}p_{i\bullet k}$, (but not between them and the row variable) then this does not imply that the column and tube components can be separated since, using Eq. (7.3),

$$\lambda_{pqr} = \sum_{i=1}^{I}\sum_{j=1}^{J}\sum_{k=1}^{K} p_{ijk}a_{ip}b_{jq}c_{kr}$$

$$= \sum_{i=1}^{I}\sum_{j=1}^{J}\sum_{k=1}^{K} p_{ij\bullet}p_{i\bullet k}a_{ip}b_{jq}c_{kr}$$

$$= \sum_{i=1}^{I} a_{ip}\left(\sum_{j=1}^{J} p_{ij\bullet}b_{jq}\right)\left(\sum_{k=1}^{K} p_{i\bullet k}c_{kr}\right)$$

$$\neq 0\,.$$

If, on the other hand, the row variable is independent of both the column and tube variables so that $p_{ijk} = p_{i\bullet\bullet}p_{\bullet jk}$ then

$$\lambda_{pqr} = \sum_{i=1}^{I}\sum_{j=1}^{J}\sum_{k=1}^{K} p_{i\bullet\bullet}p_{\bullet jk}a_{ip}b_{jq}c_{kr}$$

$$= \left(\sum_{i=1}^{I} p_{i\bullet\bullet}a_{ip}\right)\sum_{j=1}^{J}\sum_{k=1}^{K} p_{\bullet jk}b_{jq}c_{kr}$$

$$= 0\,.$$

Therefore, the row standard coordinates are all located at the origin of the biplot and so no meaningful visual interpretation can be made of the association between the three variables. Nor can one determine the association structure between the column and tube variables from this biplot. If one were interested in studying the association between these two variables, then one could perform a simple correspondence plot on the contingency table formed by aggregating across the row categories.

By using Eq. (7.10) and Eq. (7.11) to visually summarise the association between the three variables, the (i, j, k)th Pearson residual can be expressed by their inner product since

$$\gamma_{ijk} - 1 = \sum_{p=1}^{P} \tilde{f}_{ip}\tilde{g}_{jkp}\,.$$

If this residual is large, so that the position of the (j, k)th column-tube principal coordinate is in close proximity to the projection of the ith row standard coordinate from the origin, then there is a strong interaction between this row and column/tube pair.

We can easily show that the row standard coordinates are centred at the origin and have a variance of one along each dimension. To show this

$$E(\tilde{f}_{ip}) = \sum_{i=1}^{I}\sum_{p=1}^{P} p_{i\bullet\bullet}\tilde{f}_{ip}$$

$$= \sum_{i=1}^{I}\sum_{p=1}^{P} p_{i\bullet\bullet}a_{ip}$$

$$= \sum_{p=1}^{P}\left(\sum_{i=1}^{I} p_{i\bullet\bullet}a_{ip}\right)$$

$$= 0$$

while

$$Var(\tilde{f}_{ip}) = E(\tilde{f}_{ip}^{2}) - [E(\tilde{f}_{ip})]^{2}$$

$$= \sum_{i=1}^{I}\sum_{p=1}^{P} p_{i\bullet\bullet}\tilde{f}_{ip}^{2} - 0$$

$$= \sum_{i=1}^{I}\sum_{p=1}^{P} p_{i\bullet\bullet}a_{ip}^{2}$$

$$= \sum_{p=1}^{P}\left(\sum_{i=1}^{I} p_{i\bullet\bullet}a_{ip}^{2}\right)$$

$$= P$$

so that the variance along each dimension of the correspondence plot is one.

For the interactive column–tube principal coordinates, they are also centred at the origin of the correspondence plot and have a variance that is equivalent to the total inertia. To show this,

$$E(\tilde{g}_{jkp}) = \sum_{j=1}^{J}\sum_{k=1}^{K}\sum_{p=1}^{P} p_{\bullet j\bullet}p_{\bullet\bullet k}\tilde{g}_{jkp}$$

$$= \sum_{j=1}^{J}\sum_{k=1}^{K}\sum_{p=1}^{P}\sum_{q=1}^{Q}\sum_{r=1}^{R} p_{\bullet j\bullet}p_{\bullet\bullet k}b_{jq}c_{kr}\lambda_{pqr}$$

$$= \sum_{p=1}^{P}\sum_{q=1}^{Q}\sum_{r=1}^{R}\left(\sum_{j=1}^{J} p_{\bullet j\bullet}b_{jq}\right)\left(\sum_{k=1}^{K} p_{\bullet\bullet k}c_{kr}\right)\lambda_{pqr}$$

$$= 0$$

while

$$Var(\tilde{g}_{jkp}) = E(\tilde{g}_{jkp}^{2}) - [E(\tilde{g}_{jkp})]^{2}$$

$$= \sum_{j=1}^{J}\sum_{k=1}^{K}\sum_{p=1}^{P} p_{\bullet j\bullet}p_{\bullet\bullet k}\tilde{g}_{jkp}^{2} - 0$$

$$= \sum_{j=1}^{J}\sum_{k=1}^{K}\sum_{p=1}^{P} p_{\bullet j\bullet}p_{\bullet\bullet k}\left(\sum_{q=1}^{Q}\sum_{r=1}^{R} b_{jq}\lambda_{pqr}c_{kr}\right)^{2}$$

$$= \sum_{p=1}^{P}\sum_{q=1}^{Q}\sum_{r=1}^{R} \left(\sum_{j=1}^{J} p_{\bullet j\bullet} b_{jq}^2 \right) \left(\sum_{k=1}^{K} p_{\bullet\bullet k} c_{kr}^2 \right) \lambda_{pqr}^2$$

$$= \sum_{p=1}^{P}\sum_{q=1}^{Q}\sum_{r=1}^{R} \lambda_{pqr}^2$$

$$= \frac{X^2}{n} .$$

Therefore, the explained inertia along the pth dimension of the biplot is

$$\sum_{j=1}^{J}\sum_{k=1}^{K} p_{\bullet j\bullet} p_{\bullet\bullet k} \tilde{g}_{jkp}^2$$

and accounts for

$$100 \times \frac{\sum_{j=1}^{J}\sum_{k=1}^{K} p_{\bullet j\bullet} p_{\bullet\bullet k} \tilde{g}_{jkp}^2}{X^2/n} \%$$

of the total inertia. Similarly, the percentage contribution that the interaction between the jth column and kth tube categories make to the total inertia can be determined from

$$\frac{\sum_{p=1}^{P} p_{\bullet j\bullet} p_{\bullet\bullet k} \tilde{g}_{jkp}^2}{X^2/n} .$$

Therefore, if the (j, k)th column–tube principal coordinate is far from the origin this tells us that the interaction between the jth column category and kth tube category plays a strong role in defining the association structure between the variables of a three-way contingency table. Similarly, if these coordinates are close to the origin then they imply that such an interaction is not a dominant source of association between the row, column and tube variables.

Row Interactive Biplot
Suppose we now wish to construct a *row interactive biplot*. To do so, the ith row principal coordinate and the (j, k)th column–tube standard coordinate along the (q, r)th dimension of the biplot is defined by

$$\tilde{f}_{iqr} = \sum_{p=1}^{P} \lambda_{pqr} a_{ip} \tag{7.12}$$

$$\tilde{g}_{jkqr} = b_{jq} c_{kr} , \tag{7.13}$$

respectively, for $q = 1, 2, \ldots, Q$ and $r = 1, 2, \ldots R$. Therefore, this biplot is an optimal low-dimensional space that consists of QR dimensions.

These biplot coordinates show that the association between the column and tube variables that is assumed (or known from formal testing) is depicted as a projection from the origin to their standard coordinate defined by Eq. (7.13). For each of the row categories, they are depicted in the biplot using a single point defined by their principal coordinates of Eq. (7.12). One feature of these biplot coordinates is that each dimension reflects the interaction between the qth column component and the rth tube component. By using

Eq. (7.12) and Eq. (7.13) to visually summarise the association between the three variables, the (i, j, k)th Pearson residual can be expressed by their inner product such that

$$\gamma_{ijk} - 1 = \sum_{q=1}^{Q}\sum_{r=1}^{R}\tilde{f}_{iqr}\tilde{g}_{jkqr} \, .$$

Therefore, if this residual is large, so that the position of the ith row principal coordinate is in close proximity to the projection of the (j, k)th column-tube standard coordinate, then there is a strong interaction between this row and column/tube pair.

By following the same argument we made for the column-tube interaction biplot, it can be shown that the column-tube standard coordinate of Eq. (7.13) is centred at the origin of the biplot and has a variance of one along each dimension. We can also verify that the row principal coordinates defined by Eq. (7.12) are centred at the origin and have a variance in the biplot (consisting of the optimum number of dimensions, QR) of

$$\frac{X^2}{n} = \sum_{i=1}^{I}\sum_{q=1}^{Q}\sum_{r=1}^{R}p_{i\bullet\bullet}\tilde{f}_{iqr}^2$$

so that the ith row category contributes to

$$100 \times \frac{\sum_{q=1}^{Q}\sum_{r=1}^{R}p_{i\bullet\bullet}\tilde{f}_{iqr}^2}{X^2/n}\%$$

of the total association between the three variables. Therefore, if the ith row category is located far from the origin then that row is dominant in defining the association structure between the variables. However, if the point is located relatively close to the origin then it is not a dominant source of association.

Additional insight into the interpretation of the distance of a row principal coordinate from the projection of a column–tube standard coordinate can be made by examining the angle between this projection and the position of the row principal coordinate in the biplot. In doing so, let $\tilde{\mathbf{f}}_i = (\tilde{f}_{i11}, \tilde{f}_{i12}, \dots, \tilde{f}_{iQR})$ be the vector form of the ith row principal coordinate in the optimal correspondence plot. Similarly, let $\tilde{\mathbf{g}}_{jk} = (\tilde{g}_{jk11}, \tilde{g}_{jk12}, \dots, \tilde{g}_{jkQR})$ be the vector form of the (j, k) column–tube (interaction) standard coordinate in the biplot consisting of its optimal number of dimensions. Then, the angle, $\theta_{i,jk}$ between the projection of $\tilde{\mathbf{f}}_i$ and the position of $\tilde{\mathbf{g}}_{jk}$ is

$$\cos\theta_{i,jk} = \frac{\tilde{\mathbf{f}}_i\tilde{\mathbf{g}}_{jk}}{||\tilde{\mathbf{f}}_i||^2 \bullet ||\tilde{\mathbf{g}}_{jk}||^2}$$

where

$$||\tilde{\mathbf{g}}_{jk}||^2 = \sum_{q=1}^{Q}\sum_{r=1}^{R}\tilde{g}_{jkqr}^2$$

and

$$||\tilde{\mathbf{f}}_i||^2 = \sum_{q=1}^{Q}\sum_{r=1}^{R}\tilde{f}_{iqr}^2 \, .$$

Thus, from a geometric perspective, the cosine of the angle between the ith row principal coordinate and the (j, k)th column–tube (interactive) standard coordinate is equivalent to the correlation between them.

Similar conclusions can also be obtained if one is interested in interpreting the distance of the (j, k)th column–tube principal coordinate from the projection of the ith row standard coordinate in a column–tube interactive biplot. Such results therefore provide a rationale for the following rules we will be adopting when interpreting the configuration of points in a row, or column–tube, isometric biplot:

- The distance between two column-tube points has no practical meaning in a row interactive biplot. Similarly, for a column–tube interactive biplot, there is no practical interpretation of the distance between two row points.
- When studying the symmetric association between multiple variables (as we are considering in this section) using an interactive biplot, a large inner product between a row biplot coordinate and column–tube biplot coordinate implies that there exists a strong interaction between the row category and the column–tube pair.
- For the analysis of symmetric association, the origin of the interactive biplot is the point where all of the categories are located if there exists complete independence between the row, column and tube categorical variables. Coordinates that are located (relatively) far from the origin show that their categories are dominant in defining the association structure between the variables.
- The inner product of the coordinates used in the construction of an interactive biplot can be used to fully reconstruct the original three-way contingency table and allows for a more precise interpretation of the symmetric association between the variables.

7.4.3   Column-Tube Interactive Biplot for the Alligator Data

Figure 7.1 gives the two-dimensional column–tube interactive biplot from applying a symmetric multi-way correspondence analysis to Table 1.4. Since the number of row, column and tube components were chosen so that $P = Q = R = 2$ (see Section 7.3.4) the optimal biplot from this analysis consists of $P = 2$ dimensions. Recall that the chi-squared statistic from this analysis is just the sum-of-squares of the core array elements summarised in Table 7.6 and was found to be 57.127. Therefore, the total inertia of the optimal display is $57.127/219 = 0.261$ as shown in Table 7.2.

Since the optimal biplot consists of two dimensions, Figure 7.1 provides a perfect visual summary of the association between the *Size*, *Food* and *Lake* variables of Table 1.4. Note that a beta scaling (with $\beta = 3$) has been adopted in the construction of the biplot coordinates here; see Section 3.5 for a discussion of beta scaling. Such a plot is therefore a far superior visualisation of this association than the three-dimensional correspondence plots of Figure 6.2 and Figure 6.4 or the two-dimensional plot of Figure 6.5; recall that these were constructed by performing multiple correspondence analysis using the indicator and Burt matrices of Table 1.4, and from stacking this contingency table. By using the core array elements summarised in Table 7.6 for $p = 1$, we find that the first dimension of Figure 7.1 contributes to

$$100 \times \frac{(0.126)^2 + (-4.764)^2 + (-3.570)^2 + (-0.047)^2}{57.127} = 62.070\%$$

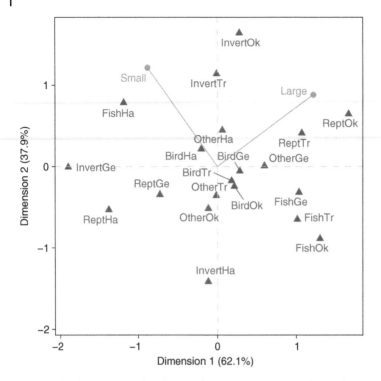

Figure 7.1 Column–tube interactive biplot from the symmetric multi-way correspondence analysis of Table 1.4 with beta scaling ($\beta = 3$).

of the total approximated inertia. Similarly the second dimension contributes to the remaining 37.93% of this inertia. This can be confirmed from the core array elements when $p = 2$ since

$$100 \times \frac{(4.599)^2 + (0.088)^2 + (0.035)^2 + (0.713)^2}{57.127} = 37.930\% \ .$$

This indeed confirms that the two dimensions of Figure 7.1 account for $62.070\% + 37.930\% = 100.000\%$ of the total approximated inertia. Both these percentage values are labeled on the axes of Figure 7.1.

For the construction of the column–tube interactive biplot of Figure 7.1, the row points (for the *Size* variable) are depicted using their standard coordinates and are calculated using Eq. (7.10). Similarly, the interaction between the column and tube categories is depicted using principal coordinates and are calculated using Eq. (7.11). These coordinates are also summarised in Table 7.8. Note that since Figure 7.1 is a column–tube biplot, the two *Size* categories are depicted as a projection from the origin to their position defined by their standard coordinate, while the interaction between the column and tube categories are depicted as points.

Figure 7.1 is therefore constructed in such a way that it assumes (or it is known) that there is a statistically significant association between the *Food* and *Lake* variables of Table 1.4. As we saw in Section 7.2.3, there is indeed a statistically significant association between these two variables and so performing a symmetric multi-way correspondence analysis that reflects the interaction between the column and tube categories is completely reasonable.

Table 7.8 Column–tube principal coordinates of the column–tube interactive biplot from the T3D of $\gamma_{ijk} - 1$ for Table 1.4

Dimension	1	2
Food/Lake	$(b_{j1}c_{k1})$	$(b_{j2}c_{k1})$
Fish/Hancock	−1.779	1.175
Invertebrate/Hancock	−0.176	−2.121
Reptile/Hancock	−2.069	−0.799
Bird/Hancock	−0.306	0.322
Other/Hancock	0.091	0.667
Fish/Oklawaha	1.948	−1.328
Invertebrate/Oklawaha	0.418	2.462
Reptile/Oklawaha	2.483	0.966
Bird/Hancock	0.323	−0.367
Other/Oklawaha	−0.173	−0.775
Fish/Trafford	1.520	−0.974
Invertebrate/Trafford	−0.017	1.708
Reptile/Trafford	1.606	0.614
Bird/Trafford	0.270	−0.264
Other/Trafford	−0.024	−0.537
Fish/George	1.553	−0.478
Invertebrate/George	−2.839	−0.010
Reptile/George	−1.091	−0.520
Bird/George	0.427	−0.084
Other/George	0.894	0.012

One of the first things to note from this biplot, however, is that not all column–tube interactions are dominant contributors to this association. For example the interaction between the food type *Bird* and those alligators in Lakes *Trafford, Hancock* and *George* is not very strong and so do not play a dominant role in defining the association structure between the three variables. This is evident by observing that their position in Figure 7.1 is located very close to the origin. On the other hand, interactions such as those between *Invertebrate* and Lake *Hancock*, *Reptile* and Lake *Hancock*, *Fish* and Lake *Oklawaha*, *Reptile* and Lake *Oklawaha*, and *Invertebrate* and Lake *Oklawaha* are very dominant contributors to the association between the three variables since they are located relatively far from the origin. Figure 7.1 also shows that not all of these interactions are strongly linked to the *Size* of the alligator. This is especially so for those column–tube interactions that are located in the bottom half of the biplot. However, the biplot does show that a strong association exists between *Small* alligators located in Lake *Hancock* that have a preference for eating *Fish*, while *Large* alligators in Lakes *Trafford* and *Oklawaha* have a preference for eating *Reptiles*.

Table 7.9 Inner products from the symmetric multi-way correspondence analysis of Table 1.4

	Small	Large
Fish/Hancock	1.996	−0.750
Invertebrate/Hancock	−1.613	−1.388
Reptile/Hancock	0.569	−2.144
Bird/Hancock	0.440	−0.058
Other/Hancock	0.486	0.466
Fish/Oklawaha	−2.219	0.797
Invertebrate/Oklawaha	1.747	1.784
Reptile/Oklawaha	−0.677	2.577
Bird/Oklawaha	−0.486	0.046
Other/Oklawaha	−0.526	−0.596
Fish/Trafford	−1.681	0.659
Invertebrate/Trafford	1.393	0.989
Reptile/Trafford	−0.446	1.660
Bird/Trafford	−0.372	0.064
Other/Trafford	−0.421	−0.334
Fish/George	−1.299	0.976
Invertebrate/George	1.660	−2.304
Reptile/George	0.220	−1.188
Bird/George	−0.319	0.296
Other/George	−0.515	0.731

The inner product of the biplot coordinates are summarised in Table 7.9. The sign and magnitude of these values helps to confirm the findings that Figure 7.1 provides. For example, recall that the close proximity of the *Small* category to the projection of the *Fish–Hancock* interaction meant that there was a strong association between *Small* alligators who eat *Fish* and are located in Lake *Hancock*. Table 7.9 shows the inner product of their points is very large and positive and has a value of 1.996. One can also see that the inner product of this column–tube pair of categories with *Large* alligators is −0.750 suggesting that a weak association exists between these alligators and those that eat *Fish* in Lake *Hancock*; note that *Fish–Hancock* is located at quite some distance from *Large* projection in Figure 7.1.

Other strong positive inner products between the column–tube interactions and alligator *Size* are summarised in Table 7.1. These include

- *Reptile/Oklawaha* and *Large* with an inner product of 2.577
- *Invertebrate/George* and *Small* with an inner product of 1.660
- *Reptile/Trafford* and *Large* with an inner product of 1.660.

Note that these strong interactions are also reflected by observing the distance of the *Food–Lake* pair from the two *Size* projections in Figure 7.1. The inner product of *Invertebrate/Oklawaha* with *Small* and *Large* sized alligators are both approximately 1.75

suggesting that there is a strong interaction between this *Food* type and *Lake* and that there is a roughly even mix of *Small* and *Large* alligators for this pair. Observing the cell frequencies of these two cells in Table 1.4 confirms that there are an equally (relatively speaking) high number of *Small* and *Large* alligators in Lake *Oklawaha* that prefer to eat *Invertebrate* animals.

There also exist very obvious interactions where the inner product suggests that there is no apparent association with the alligator *Size*. For example, the inner product of

- *Invertebrate/George* and *Large* is −2.304
- *Fish/Oklawaha* and *Small* is −2.219
- *Fish/Trafford* and *Small* is −1.681
- *Invertebrate/Hancock* and *Small* is −1.613
- *Invertebrate/Hancock* and *Large* is −1.388

and are all negative. Observing the proximity of these column–tube interactions in the biplot from the projection of the two *Size* categories in Figure 7.1 shows that the distance in each case is large reflecting that there is a very weak association between these column–tube and row combinations.

7.4.4 Row Interactive Biplot for the Alligator Data

Suppose we now construct a row interactive biplot for Table 1.4. Such a visual summary of the association is depicted by Figure 7.2. Observe that a beta scaling procedure (with $\beta = 3$)

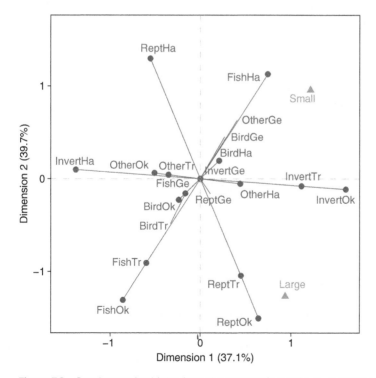

Figure 7.2 Row interactive biplot from the symmetric multi-way correspondence analysis of Table 1.4 with beta scaling ($\beta = 3$).

to the biplot coordinates has been also adopted here like in Chapter 3 Section 3.5. Since the optimal number of dimensions from a symmetric multi-way correspondence analysis of this table when $P = Q = R = 2$ consists of $QR = 2 \times 2 = 4$ dimensions, Figure 7.1 does not provide an optimal display of the association between the *Size*, *Food* and *Lake* variables. The first variable is constructed in such a way that its explained inertia reflects the interaction between the first components of the column and tube variables; that is, when $q = 1$ and $r = 1$. Therefore, by using the core array elements summarised in Table 7.6 for $q = 1$ and $r = 1$ we find that the first dimension of Figure 7.2 accounts for

$$100 \times \frac{(0.126)^2 + (4.599)^2}{57.127} = 37.052\%$$

of the total approximated inertia of the three-way contingency table; recall that the chi-squared statistic of the contingency table is 85.492 (see Table 7.1) and the total inertia is approximated since $P = Q = R = 2$. Similarly, the second dimension is constructed so that it reflects the interaction between the second column component (when $q = 2$) and the first tube component (when $r = 1$). Therefore, this dimension accounts for

$$100 \times \frac{(-4.764)^2 + (0.088)^2}{57.127} = 39.742\%$$

of the total approximated inertia. Thus, Figure 7.1 visualises 76.794% of the total inertia in Table 1.4, when $P = Q = R = 2$, and so may be considered a good visual summary of the association between the three variables. Both these percentage values are labeled on the axes of Figure 7.2. A full summary of the contribution that each of the four dimensions makes to the inertia is given in Table 7.10.

The row interactive biplot of Figure 7.2 is constructed using the row principal coordinates defined by Eq. (7.12) which are summarised in Table 7.11 while the interaction of the column-tube categories are depicted using their standard coordinates defined by Eq. (7.13). These standard coordinates are summarised in Table 7.12.

Since Figure 7.2 is a row interactive biplot, the interaction between each pair of categories from the *Food* and *Lake* variables is depicted as a projection from the origin to their standard coordinate. The two row categories of the *Size* variable are depicted in the display as points using their principal coordinate. Rather than the focus being on the interaction between the *Size* and *Lake* variables of Table 1.4 – for which, as previously described, there exists a statistically significant association – the focus now is on the difference between the *Small* alligators and the *Large* alligators. Their large distance from the origin, and from each other, indicate that there is indeed a big difference in the diet and location of these alligators and that both sizes play a very dominant role in the association structure between the three variables.

Table 7.10 Percentage of the total inertia explained by each dimension of the row interactive biplot of Table 1.4

Dimension	1	2	3	4	Total
(q, r)	(1, 1)	(2, 1)	(1, 2)	(2, 2)	
Percentage	37.052%	39.742%	22.312%	0.894%	100.000%

Table 7.11 Row principal coordinates of the row interactive biplot from the T3D of $\gamma_{ijk} - 1$ for Table 1.4

Dimension	1	2	3	4
Size	f_{i11}	f_{i21}	f_{i12}	f_{i22}
Small	3.648	2.869	2.125	0.605
Large	2.803	−3.804	−2.869	0.381

Table 7.12 Column–tube standard coordinates of the row interactive biplot from the T3D of $\gamma_{ijk} - 1$ for Table 1.4

Dimension	1	2	3	4
Food/Lake	$(b_{j1}c_{k1})$	$(b_{j2}c_{k1})$	$(b_{j1}c_{k2})$	$(b_{j2}c_{k2})$
Fish/Hancock	0.247	0.376	0.006	0.009
Invertebrate/Hancock	−0.462	0.033	−0.011	0.001
Reptile/Hancock	−0.184	0.433	−0.004	0.010
Bird/Hancock	0.068	0.065	0.002	0.002
Other/Hancock	0.145	−0.018	0.003	0.000
Fish/Oklawaha	−0.287	−0.436	0.025	0.038
Invertebrate/Oklawaha	0.536	−0.038	−0.047	0.003
Reptile/Oklawaha	0.213	−0.502	−0.019	0.044
Bird/Oklawaha	−0.079	−0.075	0.007	0.007
Other/Oklawaha	−0.169	0.021	0.015	−0.002
Fish/Trafford	−0.199	−0.303	−0.028	−0.043
Invertebrate/Trafford	0.372	−0.026	0.053	−0.004
Reptile/Trafford	0.148	−0.349	0.021	−0.050
Bird/Trafford	−0.055	−0.052	−0.008	−0.007
Other/Trafford	−0.117	0.014	−0.017	0.002
Fish/George	<0.001	−0.001	−0.426	−0.647
Invertebrate/George	0.001	<0.001	0.796	−0.057
Reptile/George	<0.001	−0.001	0.316	−0.746
Bird/George	<0.001	<0.001	−0.118	−0.112
Other/George	<0.001	<0.001	−0.251	0.031

The conclusions we made when describing the column–tube interactive biplot of Figure 7.1 also apply here for the row interactive biplot, but keeping in mind that Figure 7.2 is a visual summary of about three-quarters of the association, rather than all of it. The biplot shows that *Large* alligators, which are located in the bottom-right quadrant of Figure 7.2, have a strong interaction with those in Lake *Trafford* that eat predominately *Reptiles*. Figure 7.2 also shows that *Small* alligators, which are located in the top-right quadrant, have a strong interaction with those located in Lake *Hancock* that eat mainly *Fish*. While these highlight the dominant interactions between particular row, column and tube categories, Figure 7.2 also shows where the weaker interactions exist. For example, since alligators whose diet consists of mainly *Birds* are neither particularly *Large* or *Small* and, generally, do not play a dominant role in defining the association structure between the variables of Table 1.4. This is apparent since the position of each of the points that reflect an interaction between *Bird* and the four Florida lakes is located very close to the origin. With the interaction between *Invertebrates* and Lake *Hancock* being on the left of the display, this shows that while this interaction is dominant, the alligators are neither particularly *Large* or *Small*.

7.5 The Marcotorchino Residual $\pi_{i|j,k}$ and the Partition of τ_M

7.5.1 The Marcotrochino Residual

In Section 7.2.1, we described that the Pearson residual, $\gamma_{ijk} - 1$ of Eq. (7.1), is appropriate for assessing the departure of the (i, j, k)th cell from independence for three symmetrically associated variables. When there exists an asymmetric association structure between these residuals we can instead rely on the Marcotorchino residual

$$\pi_{i|j,k} = \frac{p_{ijk}}{p_{\bullet j \bullet} p_{\bullet \bullet k}} - p_{i \bullet \bullet} .$$

Such an index treats the row variable as being the response variable given the column and tube variables which are treated as being predictor variables. However, this index also hypothesises that the two predictor variables are independent, something that may or may not be the case. The Marcotorchino index τ_M, defined by Eq. (1.9), may be expressed in terms of these residuals such that

$$\tau_M = \frac{\sum_{i=1}^{I} \sum_{j=1}^{J} \sum_{k=1}^{K} p_{\bullet j \bullet} p_{\bullet \bullet k} \pi_{i|j,k}^2}{1 - \sum_{i=1}^{I} p_{i \bullet \bullet}^2} .$$

Therefore, if there is zero predictability of the row categories given the column and tube categories, so that each $\pi_{i|j,k} = 0$ for all $i = 1, \ldots, I, j = 1, \ldots, J$ and $k = 1, \ldots, K$, then $\tau_M = 0$. Note that the denominator of the Marcotorchino index does not contain any information about the association between the variables since it is expressed only in terms of the ith row marginal proportion, $p_{i \bullet \bullet}$. So, for the purposes of performing a multi-way correspondence analysis on a three-way contingency table with an asymmetric association structure, our focus will be on quantifying the predictability of the row categories using

$$\tau_{M(num)} = \sum_{i=1}^{I} \sum_{j=1}^{J} \sum_{k=1}^{K} p_{\bullet j \bullet} p_{\bullet \bullet k} \pi_{i|j,k}^2$$

which is the numerator of the Marcotorchino index. This index can then be used to define the total inertia of the contingency table when a non-symmetrical multi-way correspondence analysis is applied.

7.5.2 The Partition of τ_M

Like we discussed for the three-way Pearson chi-square statistic X^2, Marcotorchino's index τ_M can be partitioned to reveal further details about the structure of the asymmetric association; see, for example, Lombardo et al. (1996) and Beh et al. (2007). In doing so, we can partition Eq. (1.9) to obtain three two-way interaction terms that assess the association structure between each pair of variables, and a fourth term that reflects the trivariate association. This partition is of the form

$$\tau_M = \tau_{IJ} + \tau_{IK} + X_{JK}^2 + \tau_{IJK}$$

where

- τ_{IJ} is the Goodman–Kruskal tau index for the row (response) variable and the column (predictor) variable.
- τ_{IK} is the Goodman–Kruskal tau index for the row (response) variable and the tube (predictor) variable.
- X_{JK}^2 is Pearson's chi-squared statistic for the two (column and tube) predictor variables. This term can be used to determine whether there is any justification for assuming that there exists complete independence between these two variables.
- τ_{IJK} is that part of Marcotorchino's index that assesses the three-way asymmetric association between the three variables.

The statistical significance of these terms can be determined by rescaling Marcotorchino's index so that its C-statistic is partitioned. That is

$$C_M = C_{M_{IJ}} + C_{M_{IK}} + C_{M_{JK}} + C_{M_{IJK}}$$
$$= (n-1)(I-1)\tau_{IJ} + (n-1)(I-1)\tau_{IK}$$
$$+ \frac{(n-1)(I-1)X_{JK}^2}{1 - \sum_{i=1}^{I} p_{i\bullet\bullet}^2} + \frac{(n-1)(I-1)X_{IJK}^2}{1 - \sum_{i=1}^{I} p_{i\bullet\bullet}^2} \, .$$

Therefore, the statistical significance of the asymmetric association between the row (response) and column (predictor) variables can be made by comparing $C_{M_{IJ}}$ with the $1 - \alpha$ percentile of the chi-squared distribution with $(I-1)(J-1)$ degrees of freedom. Similarly, the statistical significance of the asymmetric association between the row (response) and tube (predictor) variables can be made by comparing $C_{M_{IK}}$ with the $1 - \alpha$ percentile of the chi-squared distribution with $(I-1)(K-1)$ degrees of freedom. The symmetric association between the column (predictor) and tube (predictor) variables can be formally tested using X_{JK}^2 and comparing it with the $1 - \alpha$ percentile of the chi-squared statistic with $(J-1)(K-1)$ degrees of freedom. The statistical significance of the trivariate association term between the three asymmetric variables can be determined by quantifying $C_{M_{IJK}}$ and comparing it with the $1 - \alpha$ percentile of the chi-squared distribution with $(I-1)(J-1)(K-1)$ degrees of freedom.

Since the total inertia for non-symmetrical multi-way correspondence analysis can be quantified by $\tau_{M(num)}$, this statistic can also be partitioned to reveal three pairwise association terms and a trivariate association term. We now describe the partition of Marcotorchino's index, its C-statistic and the total inertia, $\tau_{M(num)}$, for the alligator data summarised in Table 1.4.

7.5.3 Partition of τ_M for the Alligator Data

In Section 1.7.6 we showed that the Marcotorchino index of Table 1.4 was $\tau_M = 0.390$ with a C-statistic of $C_M = 85.120$. With a p-value that is less than 0.001 we concluded that the *Food* of preference and *Lake* in which Floridian alligators reside were statistically significant predictors of their *Size*. However, these results provide no insight into whether only *Food* or *Lake* are statistically significant predictors, or if they both are. While it is assumed that these predictor variables are independent, there is no evidence to suggest that this is indeed the case. Nor does Marcotorchino's index, or its C-statistic, tell us whether there is truly any three-way association between the variables. Therefore, we can partition Marcotorchino's index and its C-statistic into four terms; three of these terms provide insight into the association between each pair of variables while a fourth term tells us whether there exists a three-way asymmetric association. For each of these terms, we can determine their numerator which provides a partition of the total inertia, $\tau_{M(num)}$.

Table 7.13 shows that the *Food* variable is a statistically significant predictor of the *Size* variable. This is because its C-statistic of 14.705 has a p-value of 0.005. Similarly, with a p-value of 0.004, we can conclude that the *Lake* in which the alligator resides is also a statistically significant contributor to their *Size*. While Marcotorchino's index is constructed by assuming that *Food* and *Lake* are independent this is not the case; the C-statistic for the association between the column and tube variables is 38.227 and has a p-value that is less than 0.001. While there is evidence of a pair-wise association between the three variables of Table 1.4, Table 7.13 shows that the C-statistic of 18.700 for the trivariate association term has a p-value of 0.096. Thus, if one were to evaluate the association at the 0.05 level of significance, they would find that no statistically significant asymmetric association exists between the *Size*, *Food* and *Lake*. So concluding that there is an association between all three variables based solely on Marcotorchino's index, or its C-statistic, of Table 1.4 would lead to potentially misleading conclusions about the nature of this association.

Table 7.13 Partition of τ_M, C_M and τ_{Mnum} for Table 1.4

	τ_{IJ}	τ_{IK}	τ_{JK}	τ_{IJK}	**Total**
τ_M	0.067	0.062	0.175	0.086	0.390
C_M	14.705	13.489	38.227	18.700	85.120
τ_{Mnum}	0.033	0.030	0.086	0.042	0.192
%	17.275	15.847	44.909	21.969	100.000
df	4	3	12	12	31
p-value	0.005	0.004	< 0.001	0.096	< 0.001

To gain a deeper understanding into how the column and tube predictor variables are associated and how they impact upon the row variable, we can visually summarise this association using multi-way correspondence analysis. Since the association structure here is asymmetric in nature we refer to this variant as *non-symmetrical multi-way correspondence analysis*. We shall now turn our attention to discussing the foundations of this technique.

7.6 Non-symmetrical Multi-way Correspondence Analysis

7.6.1 Tucker3 Decomposition of $\pi_{i|j,k}$

In Section 7.3.1 we described how the Tucker3 decomposition (T3D) can be applied to the Pearson residuals of a three-way contingency table when performing a multi-way correspondence analysis. Such an approach assumes that the association structure between the three categorical variables of this table are symmetrically associated. However, in the case where at least one of these variables is treated as a predictor variable and at least one is treated as a predictor variable the decomposition of the Pearson residual is not appropriate. One may instead apply a T3D to the Marcotorchino residual such that

$$\pi_{i|j,k} = \sum_{p=1}^{P}\sum_{q=1}^{Q}\sum_{r=1}^{R}\lambda_{pqr}a_{ip}b_{jq}c_{kr} + e_{ijk} \tag{7.14}$$

and may be viewed as a three-way generalisation of the GSVD applied to the Goodman–Kruskal residual, $\pi_{i|j}$; see Eq. (3.2) for this decomposition. This decomposition lies at the heart of *non-symmetrical multi-way correspondence analysis* which is performed for visualising the asymmetric association between the variables. For Eq. (7.14), P, Q and R are the fixed number of components needed to visualise the asymmetric association between the I rows, J columns and K tubes. Like that of the T3D of the Pearson residuals, here the choice of P, Q and R is based on maximising the total inertia according to some minimum criterion. The a_{ip} value in Eq. (7.14) is the score of the ith row category along the pth dimension of the resulting correspondence plot for $i = 1, 2, \ldots, I$ and $p = 1, 2, \ldots, P$ so that they are constrained by

$$\sum_{i=1}^{I}a_{ip} = 0, \quad \sum_{i=1}^{I}a_{ip}^{2} = 1 \tag{7.15}$$

for each of the P components. Similarly, the b_{jq} value from the T3D of the Marcotorchino residuals is the score for the jth column category along the qth dimension for $j = 1, 2, \ldots, J$ and $q = 1, 2, \ldots, Q$ and is constrained by

$$\sum_{j=1}^{J}p_{\bullet j \bullet}b_{jq} = 0, \quad \sum_{j=1}^{J}p_{\bullet j \bullet}b_{jq}^{2} = 1$$

for the Q components. Finally, c_{kr} is the score of the kth tube category along the rth dimension for $j = 1, 2, \ldots, J$ and $r = 1, 2, \ldots, R \leq K$. These scores are constrained so that

$$\sum_{k=1}^{K}p_{\bullet\bullet k}c_{kr} = 0, \quad \sum_{k=1}^{K}p_{\bullet\bullet k}c_{kr}^{2} = 1$$

for each of the R tube components.

Note that these constraints imply that the expectation and variance of each set of scores along each dimension of the correspondence plot is zero and one, respectively.

The λ_{pqr} term in Eq. (7.14) can be interpreted as a three-way generalisation of the singular value obtained from the SVD of $\pi_{i|j}$ – see Eq. (3.6) – and is the $(p,\ q,\ r)$th element of the core array from the T3D of $\pi_{i|j,k}$. It is defined so that

$$\lambda_{pqr} = \sum_{i=1}^{I} \sum_{j=1}^{J} \sum_{k=1}^{K} p_{\bullet j\bullet} p_{\bullet\bullet k} \pi_{i|j,k} a_{ip} b_{jq} c_{kr}$$

$$= \sum_{i=1}^{I} \sum_{j=1}^{J} \sum_{k=1}^{K} p_{ijk} a_{ip} b_{jq} c_{kr} .$$

When there is zero predictability of the row categories given the column and tube categories then $p_{ijk}/(p_{\bullet j\bullet} p_{\bullet\bullet k}) = p_{i\bullet\bullet}$ or, equivalently, $p_{ijk} = p_{i\bullet\bullet} p_{\bullet j\bullet} p_{\bullet\bullet k}$ which, for all $i = 1, \ldots, I$, $j = 1, \ldots, J$ and $k = 1, \ldots, K$ implies complete independence between the three variables. In this case $\lambda_{pqr} = 0$ for $p = 1, \ldots, P$, $q = 1, \ldots, Q$ and $r = 1, \ldots, R$.

The last term of Eq. (7.14), e_{ijk}, is the error term between the observed value of $\pi_{i|j,k}$ and its reconstituted value (for some value of P, Q and R) obtained using Eq. (7.14). Just as we saw in our discussion of symmetric association, here perfect reconstitution of $\pi_{i|j,k}$ will arise when $P = I$, $Q = J$ and $R = K$. However, if $P < I$, $Q < J$ and/or $R < K$ then the error terms will not be zero and the aim is to determine the solution to a_{ip}, b_{jq} and c_{kr} so that the following error sum-of-squares function is minimised

$$SSE = \sum_{i=1}^{I} \sum_{j=1}^{J} \sum_{k=1}^{K} p_{\bullet j\bullet} p_{\bullet\bullet k} (\pi_{i|j,k} - \hat{\pi}_{i|j,k})^2$$

where

$$\hat{\pi}_{i|j,k} = \sum_{p=1}^{P} \sum_{q=1}^{Q} \sum_{r=1}^{R} \lambda_{pqr} a_{ip} b_{jq} c_{kr} .$$

In general, SSE can be minimised using an alternating least squares algorithm to obtain P, Q, R so that the solution to a_{ip}, b_{jq} and c_{kr} satisfies a given threshold of the total inertia; say 70%.

An interesting feature of the λ_{pqr} values is that their sum-of-squares can be shown to be equivalent to

$$\tau_{M(\text{num})} = \sum_{p=1}^{P} \sum_{q=1}^{Q} \sum_{r=1}^{R} \lambda_{pqr}^2 \tag{7.16}$$

when $P = I$, $Q = J$ and $R = K$. When $P < I$, $Q < J$ and/or $R < K$, then calculating $\tau_{M(\text{num})}$ using Eq. (7.16) will approximate its true value and so we can determine how similar or not this statistic is in terms of a percentage. That is, we may specify that an acceptable value of P, Q and R is one where at least 70% of $\tau_{M(\text{num})}$ is explained by the first P row components, Q column components and K tube components. Our criteria for choosing P, Q and R will be the same as the criteria we used for the multi-way correspondence analysis of symmetrically associated variables.

In the next section we express the total inertia in terms of the C_M-statistic since it can be used to formally assess the statistical significance of the asymmetric association. Therefore, the C_M-statistic can be expressed as

$$C_M = \sum_{p=1}^{P}\sum_{q=1}^{Q}\sum_{r=1}^{R}\tilde{\lambda}_{pqr}^2$$

where the weighted element of the (p, q, r)th core array is

$$\tilde{\lambda}_{pqr} = \lambda_{pqr}\sqrt{\frac{(n-1)(I-1)}{1 - \sum_{i=1}^{2}p_{i\bullet\bullet}^2}} \ .$$

7.6.2 Tucker3 Decomposition of $\pi_{i|j,k}$ and the Alligator Data

Before discussing how to visualise the asymmetric association between the variables of Table 1.4 using non-symmetrical multi-way correspondence analysis, we apply the T3D to the Marcotorchino residuals, $\pi_{i|j,k}$ of the contingency table. To do so, we first need to determine the value of P, Q and R that will satisfy our 70% threshold for the explained inertia. Table 7.14 summarises five combinations of these values, the explained inertia and the percentage of $C_M = 85.120$ that they account for. The number of iterations that it took for the least-squares procedure to converge to three decimal places is also summarised. It shows that selecting $P = Q = R = 2$ accounts for nearly 70% of the total inertia of Table 1.4. These values suggest that confining our attention to visually summarising the association using only two dimensions will provide an adequate display of the asymmetric association. For reasons of simplicity, we will describe the construction and interpretation of the row and column-tube interactive biplots with $P = Q = R = 2$ but we are aware that a further 10% can be accounted for by specifying that $Q = 3$.

With $P = 2$, $Q = 2$ and $R = 2$, we are now in a position to determine the component values of a_{ip}, b_{jq} and c_{kr} for $p = 1$, 2, $q = 1$, 2 and $r = 1$, 2. We can also calculate the weighted elements of the core array, $\tilde{\lambda}_{pqr}$ and these are summarised in Table 7.15. In fact, the sum-of-squares of these values shows that $P = Q = R = 2$ accounts for 67.41% of the total inertia quantified by $C_M = 85.120$ since

$$\sum_{p=1}^{2}\sum_{q=1}^{2}\sum_{r=1}^{2}\tilde{\lambda}_{pqr}^2 = (-2.907)^2 + (-4.021)^2 + \dots + (0.973)^2$$

Table 7.14 Inertia and percentage of explained total inertia for four combinations of P, Q and R for Table 1.4

P	Q	R	Inertia	% Explained Inertia	Iterations
1	1	1	0.069	35.94%	16
1	2	2	0.055	28.65%	33
2	2	2	0.129	67.19%	9
2	3	2	0.148	67.08%	20
2	5	4	0.192	100.00%	1

Table 7.15 The weighted core array elements, $\tilde{\lambda}_{pqr}$ from the T3D of $\pi_{i|j,k}$ of Table 1.4

	c_{k1}	
	b_{j1}	b_{j2}
a_{i1}	−2.907	−4.021
a_{i2}	4.021	−1.964

	c_{k2}	
	b_{j1}	b_{j2}
a_{i1}	−3.182	0.913
a_{i2}	−0.912	0.973

$$= 57.382$$

so that

$$100 \times \frac{57.382}{85.120} = 67.41\% \ .$$

The weighted core array elements, $\tilde{\lambda}_{pqr}$, that are summarised in Table 7.15 show that there are two $\tilde{\lambda}_{pqr}$ values that are equally the most dominant. They are $\tilde{\lambda}_{211} = 4.021$ for the interaction term $a_{i2}b_{j1}c_{k1}$ of the partition of $\pi_{i|j,k}$ while $\tilde{\lambda}_{121} = -4.021$ for the interaction term $a_{i1}b_{j2}c_{k1}$.

7.7 Constructing a Low-Dimensional Display

7.7.1 On the Choice of Coordinates

The row, column and tube component values may be used as the standard coordinates of the row, column and tube points in a visual display of the asymmetric association between the three variables. That is, (a_{i1}, a_{i2}), (b_{j1}, b_{j2}) and (c_{k1}, c_{k2}) may be used as the coordinates in a two-dimensional space to jointly depict the asymmetric association between the ith row, jth column and kth tube categories. Although, as we have discussed throughout this book, on their own, standard coordinates are not very useful in providing a visual summary of the association. Nor is it a simple task to use them for defining principal coordinates like we did when we described the difficulties concerned with the joint depiction of Eqs. (7.7)–(7.9).

Therefore, rather than jointly displaying the association between the row, column and tube categories using only their standard (or principal) coordinates, we instead use them for the construction of an interactive biplot. Since the asymmetric association of Table 1.4 is assumed to be structured so that the *Food* and *Lake* variables are the predictor variables and the *Size* variable is the response variable, we confine our attention to the construction of a column–tube interactive biplot. In this case, the interaction between the categories of the *Food* and *Lake* variables will be depicted using their principal coordinates defined by Eq. (7.11) while the *Size* categories will be visualised as a projection from the origin to their standard coordinate defined by Eq. (7.10). Constructing this column–tube interactive biplot will therefore help to determine how an alligators' type of *Food* and the *Lake* in which they are located predicts the *Size* of the alligator. We keep in mind though that the T3D is applied to $\pi_{i|j,k}$ (not $\gamma_{ijk} - 1$) and the a_{ip} values are constrained by Eq. (7.15).

Note that if the asymmetric association was structured in such a way that the tube variable was treated as the predictor variable and the row and column variables were treated as the response variables then, rather than considering the Marcotorchino index, we could instead use the delta index defined by Eq. (1.10) as the measure of association. In this case, a non-symmetrical multi-way correspondence analysis of the three-way contingency table may be performed by constructing a tube interactive biplot. In doing so, the row–column (response) interaction is depicted as a projection from the origin to their standard coordinate. The tube (predictor) categories would then be depicted as a single point at their standard coordinate. Determining how well a tube category predicts a row–column pair would then be made by observing how close (or not) a tube principal coordinate was located from the row–column projection. For example, for the analysis of Table 1.4, such a biplot would allow one to determine how the *Lake* in which an alligator is found impacts upon their *Size* and the type of *Food* they like to eat by observing the distance of a *Lake* principal coordinate from the projection made by the interaction between a *Size*–*Food* category pair.

7.7.2 Column–Tube Interactive Biplot for the Alligator Data

Performing a non-symmetrical multi-way correspondence analysis on the alligator data summarised in Table 1.4 yields the optimal column–tube interactive biplot of Figure 7.3. The optimality of this biplot is apparent since there are $P = 2$ components chosen to best represent the asymmetric association between the three-variables. Using the weighted core array values summarised in Table 7.15 we find that the explained inertia of the first dimension $(p = 1)$ is

$$(-2.907)^2 + (-4.021)^2 + (-3.182)^2 + (0.913)^2 = 35.578$$

and accounts for $100 \times 35.578/57.383 = 62.001\%$ of the total approximated inertia; the total inertia is only an approximation since $P = Q = R = 2$. Similarly, the explained inertia of the second dimension $(p = 2)$ of Figure 7.3 is

$$(4.021)^2 + (-1.964)^2 + (-0.912)^2 + (0.973)^2 = 21.804$$

and accounts for $100 \times 21.804/57.383 = 37.999\%$ of the total approximated inertia. Therefore, Figure 7.3 visually summarises 100% of the asymmetric association between the variables of Table 1.4 for $P = Q = R = 2$. Both these percentages are labeled in Figure 7.3.

This biplot is constructed by jointly depicting the projection of the row standard coordinates of the *Size* categories and the *Food-Lake* (column-tube) principal coordinates. Note that beta scaling (with $\beta = 1.5$) has been used to calculate the biplot coordinates. The row standard coordinates for the two-dimensions of Figure 7.3 are summarised in Table 7.16 while the column–tube principal coordinates are summarised in Table 7.17.

Figure 7.3 shows that an alligator in Lake *Hancock* with a preference for eating *Fish* is highly likely to lead to that alligator being *Small* in size. On the other hand, alligators that eat *Reptiles* in Lakes *Trafford* and *Oklawaha* are more likely to be *Large* in size. There

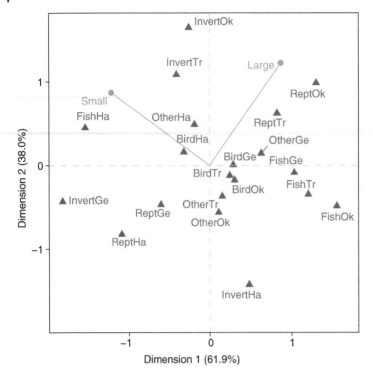

Figure 7.3 Column–tube interactive biplot of three-way non-symmetrical multi-way correspondence analysis of Table 1.4 with beta scaling ($\beta = 1.5$).

Table 7.16 Standard coordinates of the row categories from three-way non-symmetrical correspondence analysis of Table 1.4

Dimension	1	2
Size	(a_{i1})	(a_{i2})
Small	−0.815	0.580
Large	0.580	0.815

is a very strong interaction between alligators in Lakes *Trafford* and *Oklawaha* who eat mainly *Invertebrate* animals, and this column–tube interaction is equally likely to lead to alligators of *Small* and *Large* size. A similar conclusion, although one that is of a weaker prediction, can also be made for those alligators in Lake *Hancock* that eat *Other* types of animals.

While there are other strong interactions between the various *Food* and *Lake* categories, most of these are weak predictors of the *Size* of the alligator. For example, there is a strong interaction between alligators in Lake *Hancock* that eat *Invertebrate* animals and yet it is a

Table 7.17 Principal coordinates of the column–tube categories from the three-way non-symmetrical correspondence analysis of Table 1.4

Dimension	1	2
Food/Lake	$(b_{j1}c_{k1})$	$(b_{j2}c_{k1})$
Fish/Hancock	−2.313	0.674
Invertebrate/Hancock	0.724	−2.136
Reptile/Hancock	−1.633	−1.233
Bird/Hancock	−0.475	0.239
Other/Hancock	−0.282	0.729
Fish/Oklawaha	2.328	−0.728
Invertebrate/Oklawaha	−0.383	2.468
Reptile/Oklawaha	1.950	1.475
Bird/Oklawaha	0.461	−0.267
Other/Oklawaha	0.168	−0.842
Fish/Trafford	1.805	−0.520
Invertebrate/Trafford	−0.609	1.625
Reptile/Trafford	1.235	0.932
Bird/Trafford	0.373	−0.183
Other/Trafford	0.235	−0.555
Fish/George	1.548	−0.129
Invertebrate/George	−2.734	−0.649
Reptile/George	−0.904	−0.703
Bird/George	0.436	0.009
Other/George	0.945	0.216

very poor predictor of alligator *Size*. This is apparent by observing that this column–tube pair lies a long way from the projections of the *Small* and *Large* categories. There are also column–tube pairs that do not have a strong interaction and these are located close to the origin of Figure 7.3; see, for example, *Bird–Hancock* and *Bird–Trafford*.

Our discussion of the predictability of a column–tube pair of categories to a row category by observing the configuration of points in Figure 7.3 can also be verified numerically by determining the inner product of these coordinates. Such an inner product can be obtained by calculating

$$\pi_{i|j,k} = \sum_{p=1}^{2} \tilde{f}_{ip}\tilde{g}_{jkp}$$

where $\tilde{f}_{ip}$ is the standard coordinate of the ith row category along the pth dimension of the column–tube isometric biplot. Similarly, $\tilde{g}_{jkp}$ is the principal coordinate of the interaction between the jth column and kth tube categories along this dimension.

Table 7.18 Inner products from the non-symmetrical multi-way correspondence analysis of Table 1.4

Food/Lake	Size	
	Small	Large
Fish/Hancock	2.276	−0.792
Invertebrate/Hancock	−1.828	−1.321
Reptile/Hancock	0.616	−1.951
Bird/Hancock	0.526	−0.081
Other/Hancock	0.653	0.431
Fish/Oklawaha	−2.319	0.757
Invertebrate/Oklawaha	1.743	1.789
Reptile/Oklawaha	−0.734	2.332
Bird/Oklawaha	−0.530	0.050
Other/Oklawaha	−0.625	−0.589
Fish/Trafford	−1.772	0.623
Invertebrate/Trafford	1.438	0.971
Reptile/Trafford	−0.466	1.475
Bird/Trafford	−0.410	0.067
Other/Trafford	−0.513	−0.316
Fish/George	−1.336	0.792
Invertebrate/George	1.851	−2.113
Reptile/George	0.329	−1.097
Bird/George	−0.350	0.260
Other/George	−0.645	0.724

By performing a non-symmetrical multi-way correspondence analysis on Table 1.4, the inner product of the coordinates from the column–tube interactive biplot are summarised in Table 7.18. Those inner product values with a large positive value indicate that a *Food–Lake* interaction are excellent predictors of the *Size* of the alligator. For example,

- *Fish/Hancock* is a strong predictor of *Small* sized alligators and has an inner product of 2.276,
- *Reptile/Oklahawa* is a stronger predictor of *Large* sized alligators with an inner product of 2.332,
- *Invertebrate/George* is a strong predictor of *Small* alligators with an inner product of 1.851,
- *Reptile/Trafford* is a strong predictor of *Large* sized alligators with an inner product of 1.475.

Table 7.18 also shows that the strong interaction between alligators who are in Lake *Oklahawa* that eat predominately *Invertebrate* animals are good strong predictors of them being *Small* and *Large* in size with an inner product of 1.743 and 1.789 respectively. The magnitude and positive sign of these inner product values provides a numerical justification of the strong and positive asymmetric association depicted in the column–tube interactive biplot of Figure 7.3. We have also shown that this biplot visualises where there is a strong interaction between a column–tube pair of categories but have weak predictability of the alligator *Size*. These weak asymmetric associations can also be numerically verified by identifying those large, but negative, inner products that are summarised in Table 7.18. For example, the following strong column–tube interactions are very weak in predicting the *Size* of the alligator:

- the strong interaction between *Fish* and Lake *Oklahawa* is a weak predictor of *Small* alligators; its inner product is −2.319,
- the strong interaction between *Invertebrate* and Lake *George* is a weak predictor of *Large* sized alligators; its inner product is −2.113,
- the strong interaction between *Fish* and Lake *Trafford* is a weak predictor of *Small* sized alligators is −1.772,
- the strong interaction between *Reptile* and Lake *Hancock* is a weak predictor of *Large* sized alligators; its inner product is −1.951.

Table 7.18 also shows that the strong interaction between *Invertebrate* and Lake *Hancock* is a very weak predictor of *Small* and *Large* sized alligators. These asymmetric associations do not mean that, for example, alligators that eat mainly *Reptile* animals in Lake *Hancock* lead to *Small* sized alligators. Instead, one may interpret this inner product to mean that alligators in this lake with this food preference are not likely to lead to *Large* alligators. Indeed, Figure 7.3 shows that the strong column-tube interaction between Lake *Hancock* and *Reptiles* is not a good predictor for either *Size* of alligator. Such a conclusion can be made by observing that the position of *Reptile-Hancock* point (which lies in the bottom left quadrant of the biplot) lies some distance away from both of the *Size* categories (that are located in the top half of the biplot). So, some care does need to be taken to make sure that one does not make misleading, or inaccurate, predictive conclusions about the asymmetric association structure between their variables by observing only the inner-product of the first two row standard and column–tube principal coordinates.

7.8 Final Comments

In this chapter we illustrated two variants of multi-way correspondence analysis focusing on the analysis of the association between three nominal variables. These two variants were referred to as *symmetric multi-way correspondence analysis* and *non-symmetrical multi-way correspondence analysis* and involve applying a T3D to the Pearson residual (for symmetrically associated variables) and the Marcotorchino residuals (for asymmetrically associated variables), respectively.

The numerical heart of these two variants is the partition of the Pearson (three-way) chi-squared statistic when the association is symmetrically structured, and of the Marcotorchino index when the variables are asymmetrically associated. Their visual heart, however, rested in our description of the interactive biplot as the suitable approach to graphically summarise the association between the variables. Furthermore, a key advantage of these variants, when compared with the traditional approach to multiple correspondence analysis, is that multi-way correspondence analysis preserves the truly multivariate association structure of the variables; recall that the three approaches to multiple correspondence analysis we described in Chapter 6 dealt with the analysis of association that was largely bivariate in nature. Multi-way correspondence analysis also has the added advantage of partitioning the association so that one can isolate specific bivariate, or higher-variate, associations that may exist between the multiple categorical variables. Further insights can be gained when using orthogonal polynomials to reflect the structure of an ordered categorical variable and using the analog of T3D described by Lombardo, Beh and Kroonenberg (2020).

This chapter completes our introductory discussion of the various ways that one can perform correspondence analysis. However, the techniques we describe certainly do not provide an exhaustive account of the issues and approaches that one may consider. Instead, this book provides only the key features of correspondence analysis for the analysis of two nominal and/or ordinal symmetrically or asymmetrically associated variables. We have also provided an introductory description of how multiple and multi-way correspondence analysis can be performed when there are many variables to be analysed. As we described in Chapter 1, there are many excellent overviews (both introductory and in-depth) of correspondence analysis and one cannot ignore their important contribution. All variants of correspondence analysis involve, in principal, the same ideas that we have explored in this book. That is, they all involve somehow visually summarising the association between the variables of a contingency table. Other measures of association may be considered for quantifying this association. One may also wish to explore the role that categorical models play in correspondence analysis, including, but not confined to, log-linear, bilinear and association models. Irrespective of how correspondence analysis is performed, central to all methods is the need for gaining an insightful and practical understanding of the association between categorical data.

References

Agresti, A. (1990). *Categorical Data Analysis*. Wiley, New York.

Agresti, A. (1997). *An Introduction to Categorical Data Analysis*. Wiley, New York.

Agresti, A. (2002). *Categorical Data Analysis (2nd ed)*. Wiley, New York.

Agresti, A. (2010). *Analysis of Ordinal Categorical Data (2nd ed)*. Wiley, New York.

Alberti, G. (2015). CAinterprTools: An R package to help interpreting correspondence analysis' results. *SoftwareX*, 1-2:26–31.

Andersson, C. and Henrion, R. (1999). A general algorithm for obtaining simple structure of core arrays in N-way PCA with application to fluorometric data. *Computational Statistics and Data Analysis*, 31:255–278.

Anscombe, F. (1973). Graphs in statistical analysis. *The American Statistician*, 27:17–21.

Aramatte, M. (2008). Histoire et préhistoire de l'analyse des données par J.P. Benzécri: un cas de généalogie rétrospective. *Electronic Journal for History of Probability and Statistics*, 4(2): 24 pages.

Beaton, D., Fatt, C., and Abdi, H. (2014). An ExPosition of multivariate analysis with the singular value decomposition in R. *Computational Statistics and Data Analysis*, 72:176–189.

Beh, E.J. (1997). Simple correspondence analysis of ordinal cross-classifications using orthogonal polynomials. *Biometrical Journal*, 39:589–613.

Beh, E.J. (1998). A comparative study of scores for correspondence analysis with ordered categories. *Biometrical Journal*, 40:413–429.

Beh, E.J. (1999). Correspondence analysis of ranked data. *Communications in Statistics - Theory and Methods*, 28:1511–1533.

Beh, E.J. (2001a). Correspondence circles for correspondence analysis using orthogonal polynomials. *Journal of Applied Mathematics and Decision Sciences*, 5(1):35–45.

Beh, E.J. (2001b). Partitioning Pearson's chi-squared statistic for singly ordered two-way contingency tables. *Australian and New Zealand Journal of Statistics*, 43:327–333.

Beh, E.J. (2004a). S-plus code for ordinal correspondence analysis. *Computational Statistics*, 19:593–612.

Beh, E.J. (2004b). Simple correspondence analysis: A bibliographic review. *International Statistical Review*, 72:257–284.

Beh, E.J. (2008). Simple correspondence analysis of nominal-ordinal contingency tables. *Journal of Applied Mathematics and Computer Sciences*, 8:1–12.

Beh, E.J. (2010). Elliptical confidence regions for simple correspondence analysis. *Journal of Statistical Planning and Inference*, 140:2582–2588.

An Introduction to Correspondence Analysis, First Edition. Eric J. Beh and Rosaria Lombardo.
© 2021 John Wiley & Sons Ltd. Published 2021 by John Wiley & Sons Ltd.

Beh, E.J. and Davy, P. (1998). Partitioning Pearson's chi-squared statistic for a completely ordered three-way contingency table. *Australian and New Zealand Journal of Statistics*, 40:465–477.

Beh, E.J. and Davy, P. (1999). Partitioning Pearson's chi-squared statistic for a partially ordered three-way contingency table. *Australian and New Zealand Journal of Statistics*, 41:233–246.

Beh, E.J. and Farver, T. (2009). An evaluation of non-iterative methods for estimating the linear-by-linear parameter of ordinal log-linear models. *Australian and New Zealand Journal of Statistics*, 51:335–352.

Beh, E.J. and Lombardo, R. (2012). A genealogy of correspondence analysis. *Australian and New Zealand Journal of Statistics*, 54:137–168.

Beh, E.J. and Lombardo, R. (2014). *Correspondence Analysis, Practice, Methods and New Strategies*. Wiley, Chichester.

Beh, E.J. and Lombardo, R. (2015). Confidence regions and approximate p-values for classical and non-symmetric correspondence analysis. *Communications in Statistics – Theory and Methods*, 44:95–114.

Beh, E.J. and Lombardo, R. (2019a). A genealogy of correspondence analysis: Part 2 – the variants. *Electronic Journal of Applied Statistical Analysis*, 12:552–603.

Beh, E.J. and Lombardo, R. (2019b). Multiple and multi-way correspondence analysis. *WIRE's Computational Statistics*, 11(5):e1464 (11 pages).

Beh, E.J., Lombardo, R., and Simonetti, B. (2011). A European perception of food using two methods of correspondence analysis. *Food Quality and Preference*, 22:226–231.

Beh, E.J. and Simonetti, B. (2011). Investigating the European perception of food using moments obtained from non-symmetrical correspondence analysis. *Journal of Statistical Planning and Inference*, 141:2953–2960.

Beh, E.J., Simonetti, B., and D'Ambra, L. (2007). Partitioning a non-symmetric measure of association for three-way contingency tables. *Journal of Multivariate Analysis*, 98:1391–1411.

Benzécri, J.-P. (1977). Histoire et préhistoire de l'analyse des données. Partie V: l'analyse des correspondances. *Cahiers de l'Analyse des Données*, 2:9–40.

Benzécri, J.-P. (1979). Sur le calcul des taux d'inertie dans l'analyse d'un questionnaire (addendum et erratum). *Cahiers de l'Analyse des Données*, 4:377–378.

Benzécri, J.-P. (1992). *Correspondence Analysis Handbook*. Marcel-Dekker.

Best, J. and Rayner, J. (1996). Nonparametric analysis for doubly ordered two-way contingency tables. *Biometrics*, 52:1153–1156.

Bishop, Y., Fienberg, S., and Holland, P. (1975). *Discrete Multivariate Analysis: Theory and Practice*. MIT Press.

Blasius, J. (1994). Correspondence analysis in social science research. In Greenacre, M. and Blasius, J., editors, *Correspondence Analysis in the Social Sciences*, pages 23–52. Academic Press, New York.

Bock, R. (1956). The selection of judges for performance testing. *Psychometrika*, 21:349–366.

Bordet, J. (1973). Etudes de Données Geophisiques. Theése de 3éme cicle, Université de Paris VI.

Box, G. (1976). Science and statistics. *Journal of the American Statistical Association*, 71:791–799.

Box, G. (1979). Robustness in the strategy of scientific model building. In Launer, R. and Wilkinson, G., editors, *Robustness in Statistics*, pages 201–236. Academic Press, New York.

Bradley, R., Katti, S., and Coons, I. (1962). Optimal scaling for ordered categories. *Psychometrika*, 27:355–374.

Bro, R. (2006). Review on multiway analysis in chemistry – 2000–2005. *Critical Reviews in Analytical Chemistry*, 36:279–293.

Bro, R. and Kiers, H. (2003). A new efficient method for determining the number of components in PARAFAC models. *Journal of Chemometrics*, 17:274–286.

Burt, C. (1950). The factorial analysis of qualitative data. *British Journal of Statistical Psychology*, 3:166–185.

Carlier, A. and Kroonenberg, P. (1996). Decompositions and biplots in three-way correspondence analysis. *Psychometrika*, 61:355–373.

Carroll, J. and Chang, J. (1970). Analysis of individual differences in multidimensional scaling via an n-way generalization of "Eckart-Young" decomposition. *Psychometrika*, 35:283–319.

Carroll, J., Green, P., and Schaffer, C. (1986). Interpoint distance comparisons in correspondence analysis. *Journal of Marketing Research*, 23:271–280.

Carroll, J., Green, P., and Schaffer, C. (1987). Comparing interpoint distances in correspondence analysis: a clarification. *Journal of Marketing Research*, 24:445–450.

Ceulemans, E. and Kiers, H. (2006). Selecting among three-mode principal component models of different types and complexities: a numerical convex hull based method. *British Journal of Mathematical and Statistical Psychology*, 59:133–150.

Chambers, J., Cleveland, W., Kleiner, B., and Tukey, P. (1983). *Graphical Methods for Data Analysis*. Chapman and Hall.

Chessel, D., Dufour, A., and Thioulouse, J. (2004). The ade4 package I: one-table methods. *R. News*, 4/1:5–10.

Ciampi, A., Marcos, A., and Limas, M. (2005). Correspondence analysis and two-way clustering. *SORT*, 29:24–42.

Clausen, S. (1998). *Applied Correspondence Analysis, An Introduction*. Sage.

Clavel, J., Nishisato, S., and Pita, A. (2014). dualScale: Dual scaling analysis of multiple choice data. https://CRAN.R-project.org/package=dualScale. Last accessed 29 September, 2020.

Cook, R. and Weisberg, S. (1999). Graphs in statistical analysis: Is the medium the message? *The American Statistician*, 53:29–37.

D'Ambra, L. and Lauro, N. (1989). Non-symmetrical correspondence analysis for three-way contingency table. In Coppi, R. and Bolasco, S., editors, *Multiway Data Analysis*, pages 301–315. Elsevier, Amsterdam.

De Falguerolles, A. (2008). L'analyse des données; before and around. *Electronic Journal for History of Probability and Statistics*, 4(2):32 pages.

De Leeuw, J. (1984). *Canonical Analysis of Categorical Data*. DSWO Press, Leiden, The Netherlands.

De Leeuw, J. and Mair, P. (2009a). Gifi methods for optimal scaling in R: The package homals. *Journal of Statistical Software*, 31(4):1–20.

De Leeuw, J. and Mair, P. (2009b). Simple and canonical correspondence analysis using the R package anacor. *Journal of Statistical Software*, 31: 18 pages.

De Leeuw, J., van Rijckevorsel, J., and van der Wouden, H. (1981). Nonlinear principal component analysis with b-splines. *Methods of Operations Research*, 33:379–393.

Delany, M. and Abercrombie, C. (1986). American alligator habits in northcentral Florida. *The Journal of Wildlife Management*, 50:348–353.

Dray, S. and Dufour, A. (2007). The ade4 package: implementing the duality diagram for ecologists. *Journal of Statistical Software*, 22(4):20 pages.

Ellis, R., Kroonenberg, P., Harch, B., and Basford, K. (2006). Analysing environmental data from the Great Barrier Reef using nonlinear principal component analysis. *Environmetrics*, 17:1–11.

Emerson, P. (1968). Numerical construction of orthogonal polynomials from a general recurrence formula. *Biometrics*, 24:696–701.

Fienberg, S. (1975). 'Perspective Canada' as a social report. *Social Indicators Research*, 2:153–174.

Fienberg, S. and Rinaldo, A. (2007). Three centuries of categorical data analysis: Log-linear models and maximum likelihood estimation. *Journal of Statistical Planning and Inference*, 137:3430–3445.

Foucart, T. (1984). *Analyse Factorielle de Tableaux Multiples*. Masson, Paris.

Friendly, M. (2000). *Visualizing Categorical Data*. SAS Institute.

Friendly, M. (2002). A brief history of the mosaic display. *Journal of Computational Graphics and Statistics*, 11:89–107.

Gabriel, K. (1971). The biplot graphic display with application to principal component analysis. *Biometrika*, 58:453–467.

Gabriel, K. (2002). Multivariate graphics. *Encyclopedia of Statistic Science*, 8:5237–5249.

Gallego, F. (1980). Un codage flou pour l'analyse des correspondence. Thése de 3éme cycle, Université de Paris VI. Laboratoire de Statistique Mathématique.

Gallo, M. and Buccianti, A. (2013). Weighted principal component analysis for compositional data: application example for the water chemistry of the Arno River (Tuscany, central Italy). *Environmetrics*, 24:269–277.

Gautier, J. and Saporta, G. (1982). About fuzzy discrimination. In Caussinus, H., Ettinger, P., and Tomassone, R., editors, *COMPSTAT 1982 - Proceedings in Computational Statistics*, pages 224–229. Physica-Verlag.

Ghermani, B., Roux, C., and Roux, M. (1977). Sur le codage des données hétérogénes présentation de deux programmes permettant de rendre homogéne des données quelconques. *Cahiers de l'Analyse des Données*, II (1):115–118.

Gifi, A. (1990). *Non-Linear Multivariate Analysis*. Wiley, Chichester.

Goodman, L. (1970). The multivariate analysis of qualitative data: Interactions among multiple classifications. *Journal of the American Statistical Association*, 65:226–256.

Goodman, L. (1996). A single general method for the analysis of cross-classified data: reconciliation and synthesis of some methods of Pearson, Yule, and Fisher, and also some methods of correspondence analysis and association analysis. *Journal of the American Statistical Association*, 91:408–428.

Goodman, L. and Kruskal, W. (1954). Measures of association for cross classifications. *Journal of the American Statistical Association*, 49:732–764.

Gower, J., le Roux, N., and Lubbe, S. (2015). Biplots: quantitative data. *WIREs Computational Statistics*, 7:42–62.

Gower, J., le Roux, N., and Lubbe, S. (2016). Biplots: qualititative data. *WIREs Computational Statistics*, 8:82–111.

Gower, J., Lubbe, S., and le Roux, N. (2011). *Understanding Biplots*. Wiley, Chichester.

Gray, L. and Williams, J. (1981). Goodman and Kruskal τ_b multiple and partial analogs. In *Proceedings of the Social Statistics Section, American Statistical Association*, volume 10, pages 50–62.

Greenacre, M. (1984). *Theory and Application of Correspondence Analysis*. London Academic Press, London.

Greenacre, M. (1988). Correspondence analysis of multivariate categorical data by weighted least squares. *Biometrika*, 75:457–467.

Greenacre, M. (1989). The Carroll–Green–Schaffer scaling in correspondence analysis: a theoretical and empirical appraisal. *Journal of Marketing Research*, 26:358–365.

Greenacre, M. (1990). Some limitations of multiple correspondence analysis. *Computational Statistics Quarterly*, 3:249–256.

Greenacre, M. (2010). *Biplots in Practice*. Fundación BBVA, Barcelona.

Greenacre, M. (2017). *Correspondence Analysis in Practice (3rd ed)*. Chapman and Hall/CRC, Barcelona.

Greenacre, M. and Blasius, J. (2006). *Multiple Correspondence Analysis and Related Methods*. Chapman and Hall/CRC.

Greenwood, P. and Nikulin, M. (1996). *A Guide to Chi-Squared Testing*. Wiley.

Guerrero, L., Claret, A., Verbeke, W., Enderli, G., Zakowska-Biemans, S., and Vanhonacker, F. (2010). Perception of traditional food products in six European countries using free word association. *Food Quality and Preference*, 21:225–233.

Guitonneau, G. and Roux, M. (1977). Sur le taxonomie de genre erodium. *Les Cahiers de l'Analyse des Données*, II (1):97–113.

Guttman, L. (1941). The quantification of a class of attributes. In Horst, P., editor, *The Prediction of Personal Adjustment*, pages 321–347. Social Research Council: New York.

Haberman, S. (1974). Log-linear models for frequency tables with ordered classifications. *Biometrics*, 30:589–600.

Hammer, Ø., Harper, D., and Ryan, P. (2001). PAST: paleontological statistics software package for education and data analysis. *Palaeontologia Electronica*, 4(1):9 pages.

Harshman, R. (1970). Foundation of the PARAFAC procedure: Models and conditions for an "explanatory" multi-modal factor analysis. *UCLA Working Papers in Phonetics*, 16:1–84.

Harshman, R. and Lundy, M. (1984a). Data preprocessing and the extended PARAFAC model. In Law, H., Snyder Jr, C., Hattie, J., and McDonald, R., editors, *Research Methods for Multimode Data Analysis*, pages 216–284. Elsevier, New York: Praeger.

Harshman, R. and Lundy, M. (1984b). The PARAFAC model for three-way factor analysis and multidimensional scaling. In Law, H., Snyder Jr, C., Hattie, J., and McDonald, R., editors, *Research Methods for Multimode Data Analysis*, pages 122–215. Elsevier, New York: Praeger.

Harshman, R. and Lundy, M. (1994). PARAFAC - parallel factor-analysis. *Computational Statistics and Data Analysis*, 18:39–72.

Heiser, W. and Meulman, J. (1994). Homogeneity analysis: exploring the distribution of variables and their nonlinear relationships. In Greenacre, M. and Blasius, J., editors, *Correspondence Analysis in the Social Sciences: Recent Developments and Applications*, pages 179–209. Academic Press.

Hill, M. (1974). Correspondence analysis: A neglected multivariate method. *Journal of the Royal Statistical Society, Series C*, 23:340–354.

Hill, M. and Gauch Jr, H. (1980). Detrended correspondence analysis: an improved ordination techniques. *Vegetatio*, 42:47–58.

Hirschfeld, H. (1935). A connection between correlation and contingency. *Mathematical Proceedings of the Cambridge Philosophical Society*, 31:520–524.

Hoffman, D. and de Leeuw, J. (1992). Interpreting multiple correspondence analysis as a multidimensional scaling method. *Marketing Letters*, 3:259–272.

Hoffman, D., de Leeuw, J., and Arjunji, R. (1994). Multiple correspondence analysis. In Bagozzi, R., editor, *Advanced Methods of Marketing Research*, pages 260–294. John Wiley and Sons, Inc.

Hoffman, D., de Leeuw, J., and Arjunji, R. (1995). Multiple correspondence analysis. In Bagozzi, R. P., editor, *Advanced Methods of Marketing Research*, pages 260–294. Blackwell.

Holmes, S. (2008). Multivariate data analysis: The French way. In Nolan, D. and Speed, T., editors, *Probability and Statistics: Essays in Honor of David A. Freedman*, pages 219–233. Institute of Mathematical Statistics.

Imrey, P., Koch, G., Stokes, M., Darroch, J., Freeman, D., and Tolley, H. (1981). Categorical data analysis: Some reflections on the log-linear model and logistic regression. part I: Historical and methodological overview. *International Statistical Review*, 49:265–283.

Iodice D'Enza, A., Groenen, P.J.F., and van de Velden, M. (2020). PowerCA: A Fast Iterative Implementation of Correspondence Analysis. In Imaizumi, T., Nakayama, A., and Yokoyama, S., (Eds.), *Advanced Studies in Behaviormetrics and Data Science (Behaviormetrics: Quantitative Approaches to Human Behavior)*, (pp. 283–296). Singapore: Springer.

Israëls, A. (1987). *Eigenvalue Techniques for Qualitative Data*. DSWO Press, Leiden.

Kateri, M. (2010). *Contingency Table Analysis: Methods and Implementation using R*. Springer.

Kendall, M. and Stuart, A. (1967). *The Advanced Theory of Statistics, Vol. 2*. Hafner.

Kiers, H. (1997). Three-mode orthomax rotation. *Psychometrika*, 62:579–598.

Kiers, H. (1998). Three-way simplimax for oblique rotation of the three-mode factor analysis core to simple structure. *Computational Statistics and Data Analysis*, 28:207–324.

Kiers, H. (2000). Towards a standardized notation and terminology in multiway analysis. *Journal of Chemometrics*, 14:105–122.

Kiers, H. and Krijnen, W. (1991). An efficient algorithm for PARAFAC of three-way data with large numbers of observation units. *Psychometrika*, 56:147–152.

Kiers, H., Kroonenberg, P., and ten Berge, J. (1992). An efficient algorithm for TUCKALS 3 on data with large numbers of observation units. *Psychometrika*, 57:415–422.

Kiers, H., ten Berge, J., and Rocci, R. (1997). Uniqueness of three-mode factor models with sparse cores: the 3×3×3 case. *Psychometrika*, 62:349–374.

Killion, R. and Zahn, D. (1976). A bibliography of contingency table literature: 1900–1974. *International Statistical Review*, 44:71–112.

Konig, R. (2010). Changing social categories in a changing society: studying trends with correspondence analysis. *Quality and Quantity*, 44:409–425.

Kroonenberg, P. (1983). *Three mode Principal Component Analysis*. DSWO Press, Leiden.

Kroonenberg, P. (1985). Three-mode principal component analysis of semantic differential data: The case of a triple personality. *Applied Psychological Measurement*, 9:83–94.

Kroonenberg, P. (1987). Multivariate and longitudinal data on growing children. solutions using a three-mode principal component analysis and some comparison results with other approaches. In Jansenn, J., Marcotorchino, F., and Proth, J., editors, *Data Analysis. The Ins and Outs of Solving Real Problems*, pages 89–112. Plenum.

Kroonenberg, P. (1989). Singular value decomposition of interactions in three-way contingency tables. In Coppi, R. and Bolasco, S., editors, *Multiway Data Analysis*, pages 169–184. Elsevier.

Kroonenberg, P. (1994). The TUCKALS line: A suite of programs for three-way data analysis. *Computational Statistics and Data Analysis*, 18:73–96.

Kroonenberg, P. (2005). Model selection procedures in three-mode component models. In Vichi, M., Molinari, P., Mignani, S., and Montanari, A., editors, *New Developments in Classification and Data Analysis*, pages 167–172. Elsevier.

Kroonenberg, P. (2008). *Applied Multiway Data Analysis*. Wiley.

Kroonenberg, P. and de Leeuw, J. (1980). Principal component analysis of three mode data by means of alternating least squares algorithms. *Psychometrika*, 45:69–97.

Kroonenberg, P. and Lombardo, R. (1999). Nonsymmetric correspondence analysis: A tool for analysing contingency tables with a dependence structure. *Multivariate Behavioral Research Journal*, 34:367–397.

Kroonenberg, P. and Oort, F. (2003). Three-mode analysis of multimode covariance matrices. *British Journal of Mathematical and Statistical Psychology*, 56:305–336.

Kroonenberg, P. and ten Berge, J. (2011). The equivalence of Tucker3 and PARAFAC models with two components. *Chemometrics and Intelligent Laboratory Systems*, 106:21–26.

Lancaster, H. (1951). Complex contingency tables treated by the partition of the chi-square. *Journal of Royal Statistical Society (Series B)*, 13:242–249.

Lancaster, H. (1953). A reconstitution of χ^2 considered from metrical and enumerative aspects. *Sankhya*, 13:1–107.

Lancaster, H. (1969). *The Chi-Squared Distribution*. John Wiley and Sons, New-York.

Lauro, N. and D'Ambra, L. (1984). L'analyse non symmétrique des correspondances. In Diday, E., editor, *Data Analysis and Informatics, III*, pages 433–446. North-Holland.

Lê, S., Josse, J., and Husson, F. (2008). FactoMineR: An R package for multivariate analysis. *Journal of Statistical Software*, 25(1):18 pages.

Le Foll, Y. (1979). Sur les propriétés de l'analyse des correspondances pour diverses formes completes de données. Thése de 3éme Cycle, Université de Paris VI.

Le Roux,B. and Rouanet, H. (2004). *Geometric Data Analysis: From Correspondence Analysis to Structured Data Analysis*. Kluwer Academic Publications.

Lebart, L. (1976). The significance of eigenvalues issued from correspondence analysis of contingency tables. In Gordesch, J. and Naeve, P., editors, *COMPSTAT1976*, pages 38–45. Physica-Verlag.

Lebart, L., Morineau, A., and Warwick, K. (1984). *Multivariate Descriptive Statistical Analysis*. Wiley, New-York.

Lebart, L., (1988). Visualizations of textual data. In Blasius, J. and Greenacre, M., editors, *Visualizations of Categorical Data*, pages 133–147. Academic Press.

Leclerc, A. (1975). L'anlyse des correspondances sur juxtaposition de tabeaux de contingence. *Revue de Statistique Appliquée*, 23(3):5–16.

Leibovici, D. (2010). Spatio-temporal multiway decomposition using principal tensor analysis on k-modes: The R package PTAk. *Journal of Statistical Software*, 34(10):34 pages.

Leibovici, D. (2015). Principal tensor analysis on k modes. https://cran.r-project.org/web/packages/PTAk/index.html. Last accessed 29 September, 2020.

Librero, A., Willems, P., and Villardon, P. (2015). cncaGUI: Canonical Non-Symmetrical Correspondence Analysis in R. https://CRAN.R-project.org/package=cncaGUI. Last accessed 29 September, 2020.

Liebetrau, A. (1983). *Measures of Association*. Sage Publications.

Light, R. and Margolin, B. (1971). An analysis of variance for categorical data. *Journal of the American Statistical Association*, 66:534–544.

Linting, M., Meulman, J., Groenen, P., and van der Kooij, A. (2007). Stability of nonlinear principal components analysis: An empirical study using the balanced bootstrap. *Psychological Methods*, 12:359–379.

Loisel, S. and Takane, Y. (2016). Partitions of Pearson's chi-square statistic for frequency tables: A comprehensive account. *Computational Statistics*, 31:1429–1452.

Lombardo, R. (2011). Three-way association measure decompositions: the delta index. *Journal of Statistical Planning and Inference*, 141:1789–1799.

Lombardo, R. and Beh, E.J. (2010). Simple and multiple correspondence analysis for ordinal-scale variables using orthogonal polynomials. *Journal of Applied Statistics*, 37:2101–2116.

Lombardo, R. and Beh, E.J. (2016). Variants of simple correspondence analysis. *The R. Journal*, 8/2:167–184.

Lombardo, R. and Beh, E.J. (2017). Three-way correspondence analysis for ordinal nominal variables. In Petrucci, A. . and Verde, R., editors, *SIS 2017 Statistics and Data Science: New Challenges, New Generations*, pages 613–620. Firenze Press.

Lombardo, R., Beh, E.J., and D'Ambra, A. (2011). Studying the dependence between ordinal-nominal categorical variables via orthogonal polynomials. *Journal of Applied Statistics*, 38:2119–2132.

Lombardo, R., Beh, E.J., and D'Ambra, L. (2007). Non-symmetric correspondence analysis with ordinal variables. *Computational Statistics and Data Analysis*, 52:566–577.

Lombardo, R., Beh, E.J., and Guerrero, L. (2019). Analysis of three-way non-symmetrical association of food concepts in cross-cultural marketing. *Quality and Quantity*, 53:2323–2337.

Lombardo, R., Beh, E.J., and Kroonenberg, P. (2016a). Modelling trends in ordered correspondence analysis using orthogonal polynomials. *Psychometrika*, 81:325–349.

Lombardo, R., Carlier, A., and D'Ambra, L. (1996). Nonsymmetric correspondence analysis for three-way contingency tables. *Methodologica*, 4:59–80.

Lombardo, R., Kroonenberg, P., and Beh, E.J. (2016b). Modelling trends in ordered three-way non-symmetrical correspondence analysis. In Pratesi, M. and Perna, C., editors, *Proceedings of the 48th Scientific Meeting of the Italian Statistical Society, June 8-10, 2016*, 14 pages. Springer.

Lombardo, R., Beh, E.J., and Kroonenberg, P.M. (2020a). Symmetrical and non-symmetrical variants of three-way correspondence analysis for ordered variables, *Statistical Science* (in press).

Lombardo, R. and Meulman, J. (2010). Multiple correspondence analysis via polynomial transformations of ordered categorical variables. *Journal of Classification*, 27:191–210.

Lombardo, R. and Ringrose, T. (2012). Bootstrap confidence regions in non-symmetrical correspondence analysis. *Electronic Journal of Applied Statistical Analysis*, 5:413–417.

Lombardo, R., Takane, Y., and Beh, E.J. (2020b). Familywise decompositions of Pearson's chi-square statistic in the analysis of contingency tables. *Advances in Data Analysis and Classification*, 14:629–649.

Lombardo, R. and van Rijckevorsel, J. (2001). Interactions terms in homogeneity analysis: Higher order non-linear multiple correspondence analysis. In Borra, S., Rocci, R., Vichi, M., and Schader, M., editors, *Advances in Classification and Data Analysis*. Springer.

Lorenza-Seva, U. (2011). Horn's parallel analysis for selecting the number of dimensions in correspondence analysis. *European Journal of Research Methods for the Behavioral and Social Sciences*, 7:96–102.

Markus, M. (1994). *Bootstrap Confidence Regions in Non-Linear Multivariate Analysis*. DSWO Press.

Martin, J. (1980). Le codage flou et ses applications en statistique. Thése de 3éme cycle, Université de Pau et des pays de l'Ardour.

Meulman, J., van der Kooij, A., and Heiser, W. (2004). Principal component analysis with nonlinear optimal scaling transformations for ordinal and nominal data. In Kaplan, D., editor, *Handbook of Quantitative Methods in the Social Sciences*, pages 49–70. CA: Sage, Newbury Park.

Murakami, T. and Kroonenberg, P. (2003). Three-mode models and individual differences in semantic differential data. *Multivariate Behavioral Research*, 38:247–283.

Murtagh, F. (2005). *Correspondence Analysis and Data Coding with Java and R*. Boca Raton, FL: Chapman and Hall/CRC.

Nair, V. (1986). Testing an industrial reduction method with ordered categorical data. *Technometrics*, 28:283–311.

Nenadić, O. and Greenacre, M. (2007). Correspondence analysis in R, with two- and three-dimensional graphics: The ca package. *Journal of Statistical Software*, 20:1–13.

Nishisato, S. (1980). *Analysis of Categorical Data: Dual Scaling and its Applications*. University of Toronto Press, Toronto.

Nishisato, S. (1988). Assessing quality of joint graphical display in correspondence analysis and dual scaling. In Diday, E., editor, *Data Analysis and Informatics, V*, pages 409–416. Elsevier Science Publishers B.V. (North-Holland).

Nishisato, S. (1994). *Elements of Dual Scaling: An Introduction to Practical Data Analysis*. Taylor and Francis Group, LLC.

Nishisato, S. (1995). Graphical representation quantified categorical data: its inherent problems. *Journal of Statistical Planning and Inference*, 43:121–132.

Nishisato, S. (2007). *Multidimensional Nonlinear Descriptive Analysis*. Taylor and Francis Group, LLC.

Nishisato, S. and Arri, P. (1975). Non-linear programming approach to optimal scaling of partially ordered categories. *Psychometrika*, 40:525–547.

Nishisato, S. and Clavel, J. (2003). A note on between-set distances in dual scaling and correspondence analysis. *Behaviormetrika*, 30:87–98.

Nishisato, S. and Clavel, J. (2010). Total information analysis: Comprehensive dual scaling. *Behaviormetrika*, 37:15–32.

Oksanen, J., Blanchet, F., Friendly, M., Kindt, R., Legendre, P., McGlinn, D., Minchin, P., O'Hara, R., Simpson, G., Solymos, P., Stevens, M., Szoecs, E., and Wagner, H. (2016). vegan:

Community ecology package. https://CRAN.R-project.org/package=vegan. Last accessed 29 September, 2020.

Parsa, A. and Smith, W. (1993). Scoring under ordered constraints in contingency tables. *Communications in Statistics – Theory and Methods*, 22:3537–3551.

Pearson, K. (1895). Note on regression and inheritance in the case of two parents. *Proceedings of the Royal Society of London*, 58:240–242.

Pearson, K. (1904). On the theory of contingency and its relation to association and normal correlation. *Drapers Memoirs*, Biometric Series, volume 1, London.

Rayner, J. and Beh, E.J. (2009). Towards a better understanding of correlation. *Statistica Neerlandica*, 63:324–333.

Rayner, J. and Best, D. (1996). Smooth extensions of Pearson's product moment correlation and Spearman's rho. *Statistics and Probability Letters*, 30:171–177.

Rayner, J. and Best, D. (2000). Analysis of singly ordered two-way contingency tables. *Journal of Applied Mathematics and Decision Sciences*, 4:83–98.

Rayner, J. and Best, D. (2001). *A Contingency Table Approach to Nonparametric Testing*. Chapman and Hall/CRC.

Ringrose, T. (1992). Bootstrapping and correspondence analysis in archaeology. *Journal of Archaeological Science*, 19:615–629.

Ringrose, T. (1996). Alternative confidence regions for canonical variate analysis. *Biometrika*, 83:575–587.

Ringrose, T. (2012). Bootstrap confidence regions for correspondence analysis. *Journal of Statistical Computation and Simulation*, 83:1397–1413.

Ripley, B. (2016). MASS: Support functions and datasets for Venables and Ripley's MASS (version 7.3-51.5). https://cran.r-project.org/web/packages/MASS/index.html. Last accessed 29 September, 2020.

Ritov, Y. and Gilula, Z. (1983). Analysis of contingency tables by correspondence models subject to order constraints. *Journal of the American Statistical Association*, 88:1380–1387.

Rocci, R. (1992). Three-mode factor analysis with binary core and orthonormality constraints. *Journal of the Italian Statistical Society*, 70:413–422.

Rocci, R. (2001). Core matrix rotation to natural zeros in three-mode factor analysis. In Borra, S., Rocci, R., Vichi, M., and Schader, M., editors, *Advances in Classification and Data Analysis: Studies in Classification, Data Analysis, and Knowledge Organization*, pages 161–168. Springer.

Rocci, R. and Vichi, M. (2005). Three-mode component analysis with crisp or fuzzy partition of units. *Psychometrika*, 70:715–736.

Schriever, B. (1983). Scaling of order dependent categorical variables with correspondence analysis. *International Statistical Review*, 51:225–238.

Simonetti, B., Beh, E.J., and D'Ambra, L. (2011). The analysis of dependence for three way contingency tables with ordinal variables: a case study of parient satisfaction. *Journal of Applied Statistics*, 37:91–103.

Smilde, A. (1992). Three-way analyses - problems and prospects. *Chemometrics and Intelligent Laboratory Systems*, 15:143–157.

Smilde, A., Bro, R., and Geladi, P. (2004a). *Multi-way Analysis: Applications in the Chemical Sciences*. Wiley.

Smilde, A., Geladi, P., and Bro, R. (2004b). *Multi-way Analysis in Chemistry*. Wiley.

Snell, E. (1964). The scaling procedure for ordered categorical data. *Biometrics*, 20:592–607.

Spain, S., Miner, A., Kroonenberg, P., and Drasgow, F. (2010). Job performance as multivariate dynamic criteria: Experience sampling and multiway component analysis. *Multivariate Behavioral Research*, 45:599–626.

Stigler, S. (2002). The missing early history of contingency tables. *Annales de la Faculté Des Sciences de Toulouse*, 11:563–573.

Takane, Y. and Jung, S. (2009a). Regularized nonsymmetric correspondence analysis. *Computational Statistics and Data Analysis*, 53:3159–3170.

Takane, Y. and Jung, S. (2009b). Tests for ignoring and eliminating in nonsymmetrical correspondence analysis. *Advances in Data Analysis and Classification*, 3:315–340.

Teil, H. (1975). Correspondence factor analysis: An outline of its method. *Mathematical Geology*, 7, 1:3–12.

Ten Berge, J. and Kiers, H. (1999). Simplicity of core arrays in three-way principal component analysis and the typical rank of p. *Linear Algebra and its Applications*, 294:9–179.

Tenenhaus, M. and Young, F. (1985). An analysis and synthesis of multiple correspondence analysis, optimal scaling, dual scaling, homogeneity analysis and other methods for quantifying categorical multivariate data. *Psychometrika*, 50:91–119.

Ter Braak, C. (1986). Canonical correspondence analysis: a new eigenvector technique for multivariate direct gradient analysis. *Ecology*, 67:1167–1179.

Theus, M. (2012). Mosaic plots. *WIREs Computational Statistics*, 4:191–198.

Thioulouse, J., Chessel, D., Doledec, S., and Olivier, J.-M. (1997). ADE-4: a multivariate analysis and graphical display software. *Statistics and Computing*, 7:75–83.

Timmerman, M. and Kiers, H. (2000). Three-mode principal component analysis: Choosing the numbers of components and sensitivity to local optima. *British Journal of Mathematical and Statistical Psychology*, 53:1–16.

Torgerson, W. (1958). *Theory and Methods of Scaling*. Wiley, New York.

Tucker, L. (1963). Implications of factor analysis of three-way matrices for measurement of change. In Harris, C., editor, *Problems in Measuring Change*, pages 122–137. The University of Wisconsin Press, Madison.

Tucker, L. (1964). The extension of factor analysis to three-dimensional matrices. In Frederiksen, N. and Gulliksen, H., editors, *Contributions to Mathematical Psychology*, pages 109–127. Holt, Rinehart and Winston, Inc.

Tucker, L. (1966). Some mathematical notes on three mode factor analysis. *Psychometrika*, 31:279–311.

Upton, G. (2000). Cobweb diagrams for multiway contingency tables. *The Statistician*, 49:79–85.

Van Herk, H. and van de Velden, M. (2007). Insight into the relative merits of rating and ranking in a cross-national context using three-way correspondence analysis. *Food Quality and Preference*, 18:1096–1105.

Van Meter,K., Schiltz, M.-A., Cibois, P., and Mounier, L. (1994). Correspondence analysis: A history and French sociological perspective. In Greenacre, M. and Blasius, J., editors, *Correspondence Analysis in the Social Sciences*, pages 128–137. Elsevier.

Van Rijckevorsel, J. (1987). *The Application of Fuzzy Coding and Horseshoes in Multiple Correspondence Analysis*. DSWO Press, Leiden.

Van Rijckevorsel, J. (1988). Fuzzy coding and B-splines. In Van Rijckevorsel, J. and de Leeuw, J., editors, *Component and Correspondence Analysis*, pages 33–54. John Wiley and Sons, Inc.

Venables, W. and Ripley, B. (2002). *Modern Applied Statistics with R* (4th ed). Springer-Verlag.

Wegman, E. and Solka, J. (2002). On some mathematics for visualizing high dimensional data. *Sankhya*, 64:429–452.

Weller, S. and Romney, A. (1990). *Metric Scaling: Correspondence Analysis.* Sage University Paper Series on Quantitative Applications in the Social Sciences 07-075. Newbury Park, CA: Sage.

Williams, O. and Grizzle, J. (1972). Analysis of contingency tables having ordered response categories. *Journal of the American Statistical Association*, 67:55–63.

Yang, K.-S. and Huh, M.-H. (1999). Correspondence analysis of two-way contingency tables with ordered column categories. *Journal of the Korean Statistical Society*, 28:347–358.

Yule, G. and Kendall, M. (1950). *An Introduction to the Theory of Statistics.* Charles Griffin.

Author Index

An Introduction to Correspondence Analysis, First Edition. Eric J. Beh and Rosaria Lombardo.
© 2021 John Wiley & Sons Ltd. Published 2021 by John Wiley & Sons Ltd.

Subject Index

An Introduction to Correspondence Analysis, First Edition. Eric J. Beh and Rosaria Lombardo.
© 2021 John Wiley & Sons Ltd. Published 2021 by John Wiley & Sons Ltd.